HOW TO DRAW ROBOTS

Written by
SHERARD JACKSON
and
ROBERT ACOSTA

Illustrated by
SHERARD JACKSON

Additional Art by
BEN DUNN

Covers Colored by
BEN HUNZEKER

Editing
ROBERT ACOSTA
DOUG DLIN

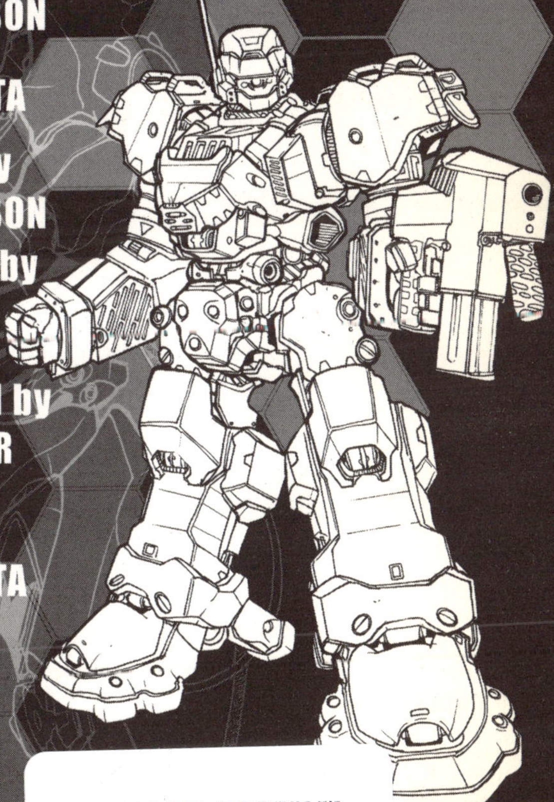

***Antarctic Press Presents: How to Draw Robots* Pocket Manga #1, April 2010**, is published by Antarctic Press, 7272 Wurzbach, Suite #204, San Antonio, Texas 78240. FAX#: (210) 593-0692. Art and text © Sherard Jackson and Robert Acosta. Additional art © Ben Dunn. All other material is ™ and © Antarctic Press. No similarity to any actual character(s) and/or place(s) is intended, and any such similarity is entirely coincidental. Nothing from this book may be reproduced without the express written consent of the authors, except for purposes of review or promotion. *"Let's discontinue this big, naughty, battle toy once and for all!"* Printed and bound in Canada.

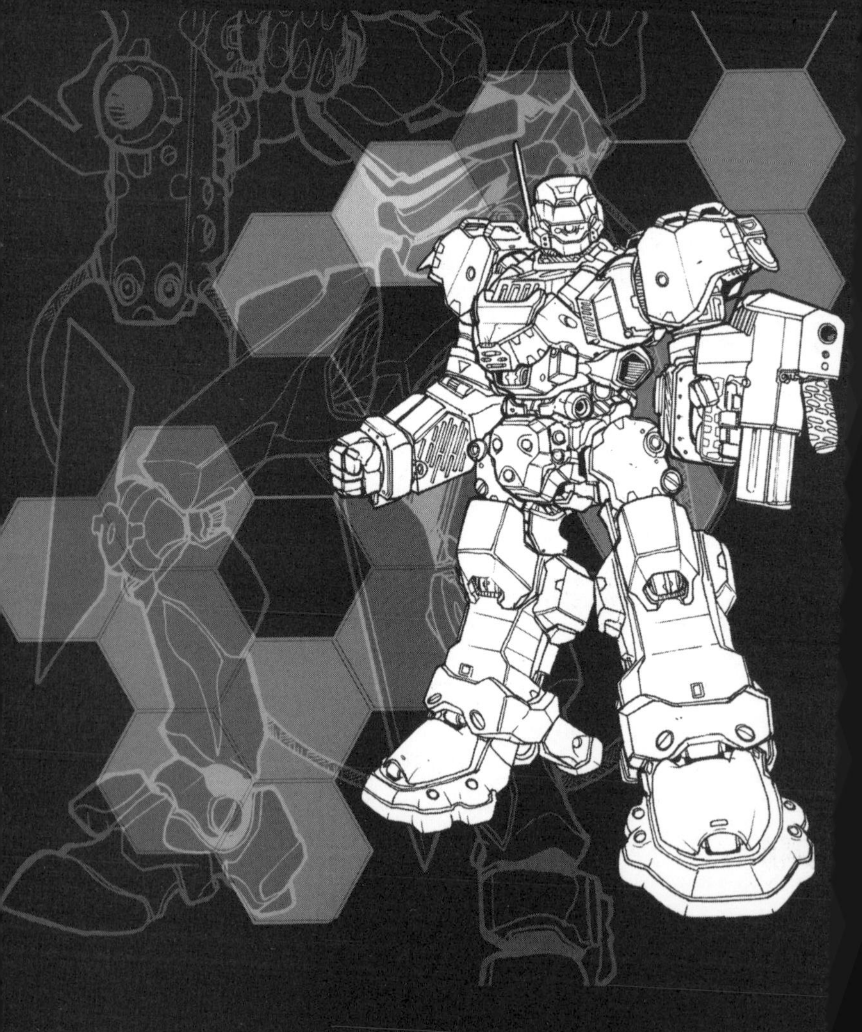

TABLE OF CONTENTS

INTRODUCTION

Welcome to *How to Draw Robots*! Contained in this book are many of the tips and tricks I have learned over the years on my way to becoming a better mecha artist. Hopefully, all these things will help you along also.

Before we start, we should go over just what "mecha" is. The word "mecha" (same for singular or plural) is actually short for the word "mechanism." In the case of Japanese anime and manga, it is also a generic term for robots or similar machines. More often than not, one of the defining things about a mecha is that it's some sort of large, piloted robot used as a tool by the operator to perform larger-than-life tasks. Those tasks could include lifting heavy objects, performing police or military duties, or even battling super-large monsters bent on terrorizing the planet. In most anime and manga stories, the mecha are not sentient but are more like cars, trucks, or tanks to the characters using them. Mecha are usually built to look like the species employing them, sort of a "creating in your own image" kinda thing.

This book is going to get you on the road to drawing better mecha. Through the instruction and examples, you'll understand the "how's" and "why's" of designing mecha as well as get to see some original ways to approach designing your very own. There's a lot of good, practical stuff in here, along with some fun stuff as well. One thing to remember, though: No matter how many "How-to" books you get or how expensive your art supplies are, that's no replacement for hard work and lots and lots of practice.

So read on, have fun, and enjoy. Your mecha education is about to begin!

-Sherard-

CHAPTER 01

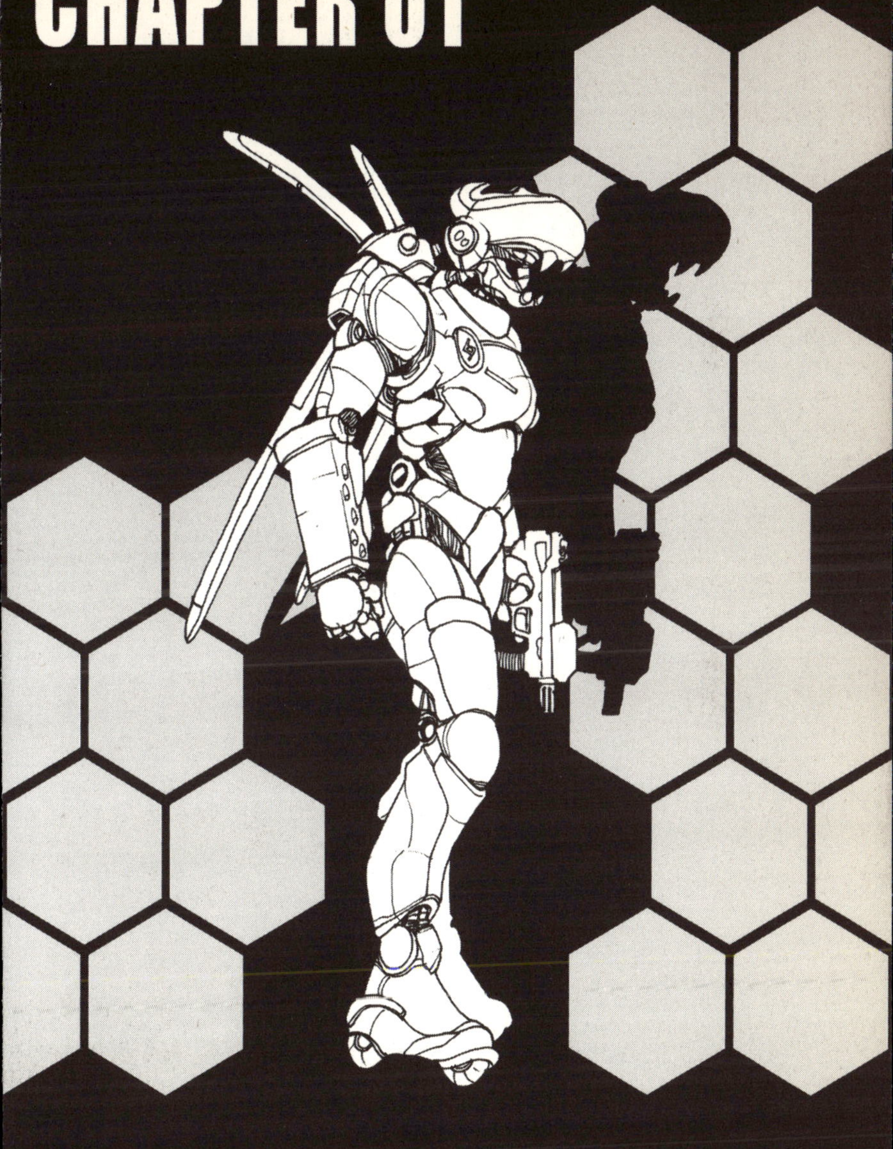

BASIC SETUPS

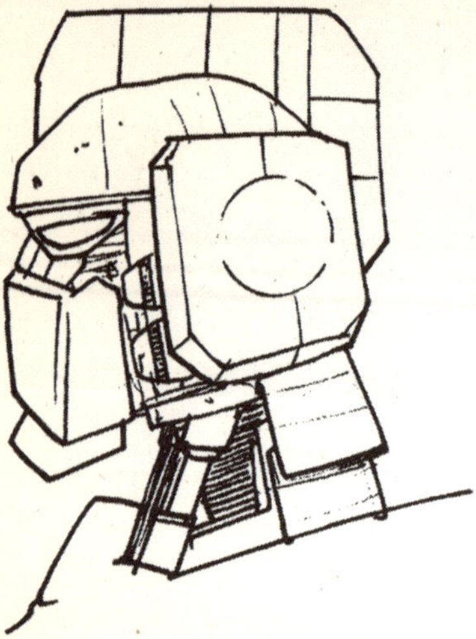

HEAD DESIGNS

When designing mecha, approach it the way you would start with any other character in your story. Here, we're going to start with the head. Mecha heads come in many different sizes and shapes. Pick one that will reflect the overall look and purpose of your mech. The function of your mech should determine the way the head and face are designed. If the mech is a public servant type, there would likely be a more human-like face on it. A labor mech would have no need for this because it would probably be working away from people.

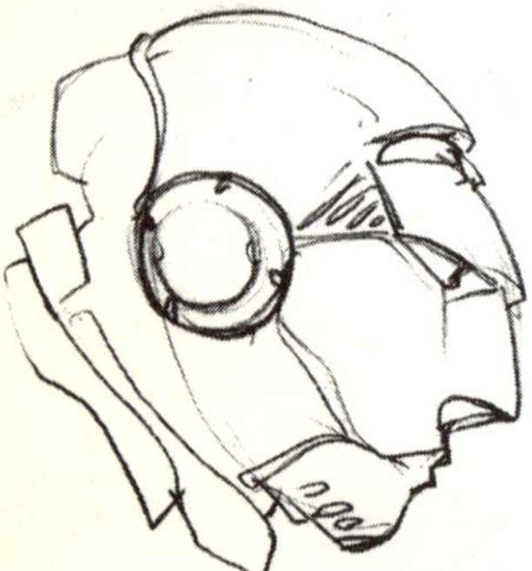

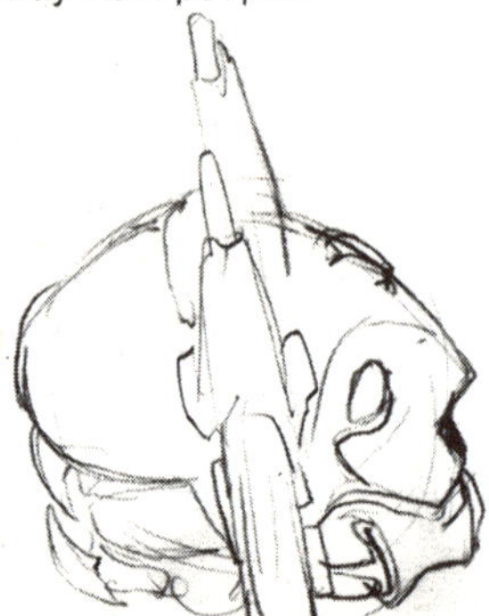

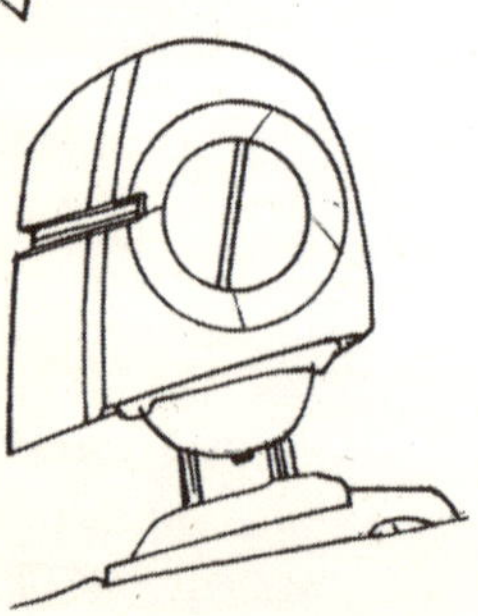

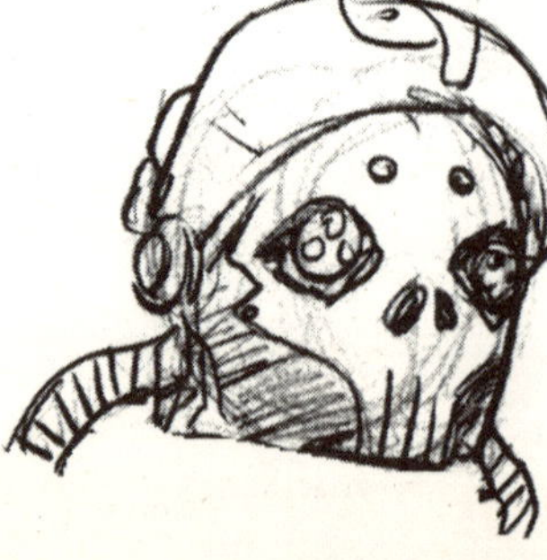

EXAMPLES OF EYES

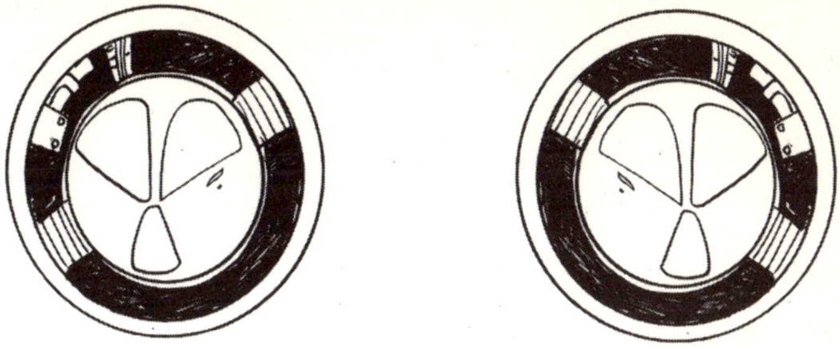

Designing the eyes can be one of the more challenging aspects of creating mechs. Because the mech is a tool operated by humans in most cases, everything in the design has to have a purpose but still fit into the humanoid form. In the case of the eyes, not only are they imporant to completing the look of your mech, but they also play the role of camera bays or cockpit windows. The more human-looking the eyes are, the easier it is for the viewer to relate to the mech sporting them.

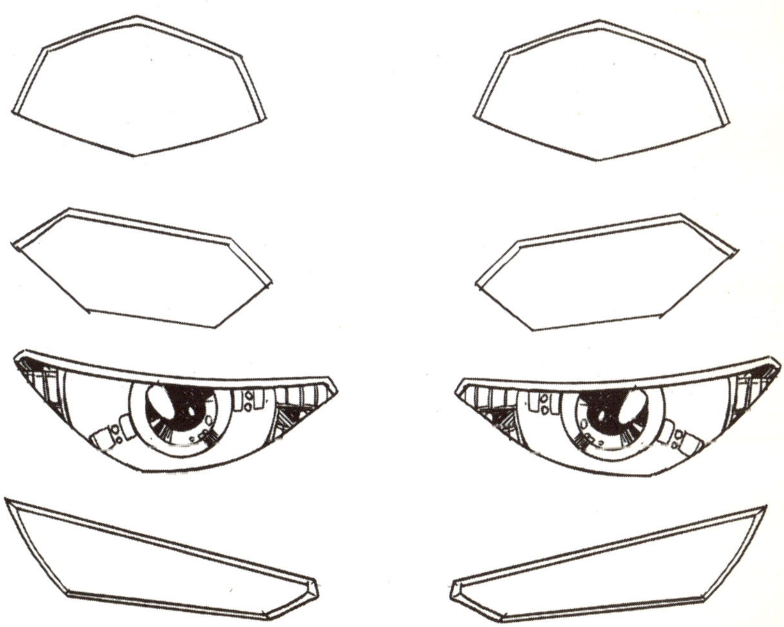

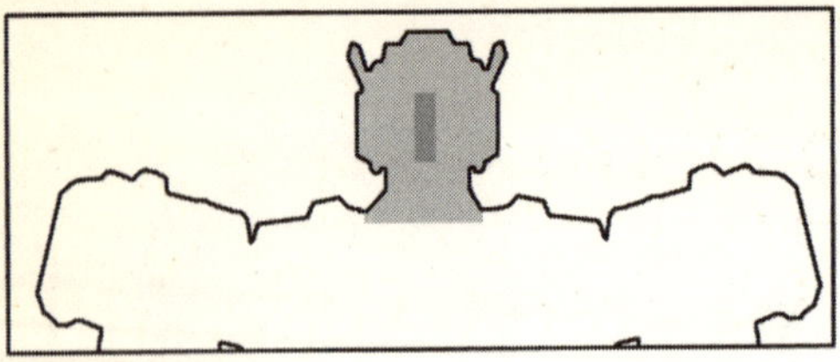

NOSES

Noses can be as characteristically distinct as eyes when you're creating your mecha. Though they may not be used for the purpose of breathing and smelling, they still help the mech's face look more interesting and more human.

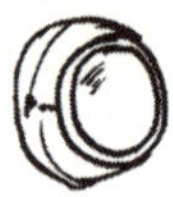

Button

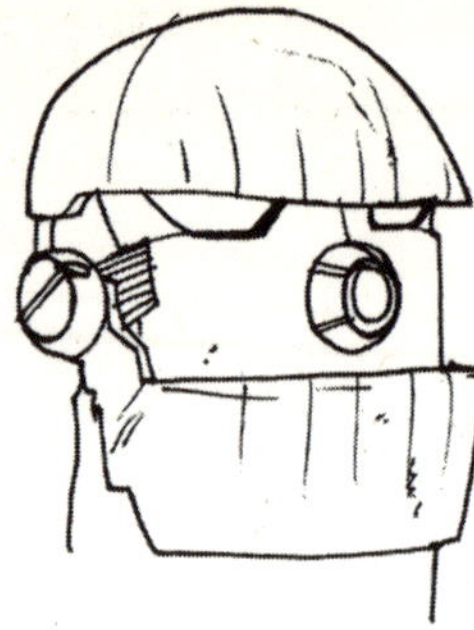

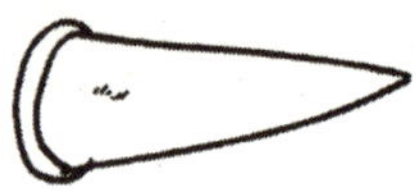

Cone

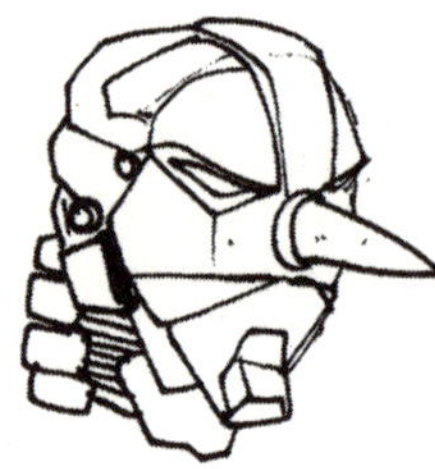

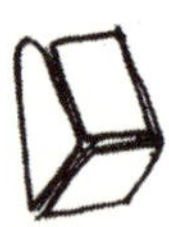

Square-shaped

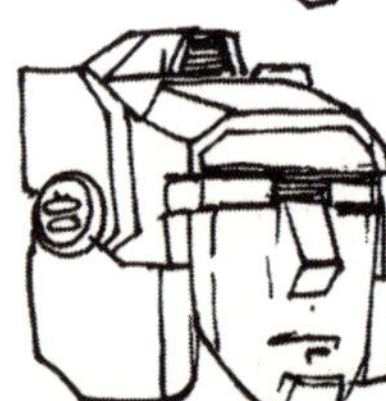

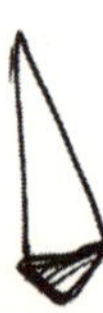

Triangle

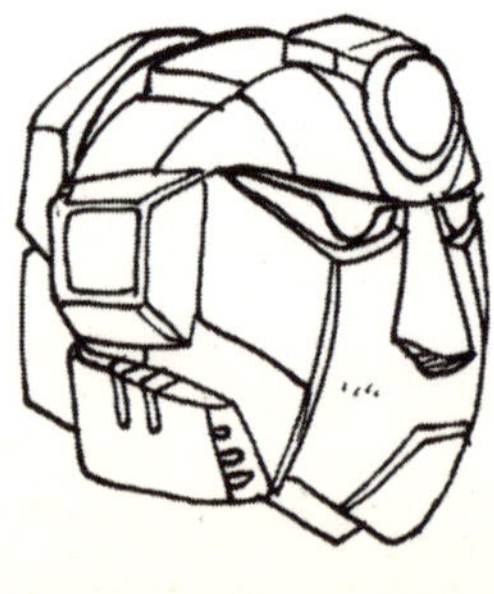

HEAD FEATURES

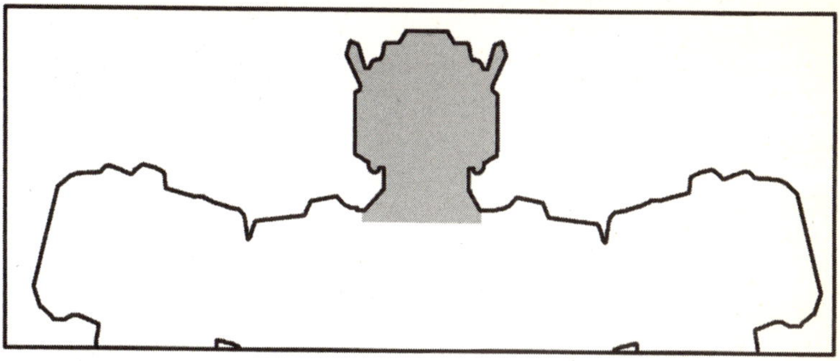

MOUTHS

In many cases, mouths aren't important. Because mechs don't usually speak, most of the time they don't even really require mouths. If you design a mech that does speak, when you design the mouth, determine what level of expressions you think you're going to want out of your mech and design accordingly.

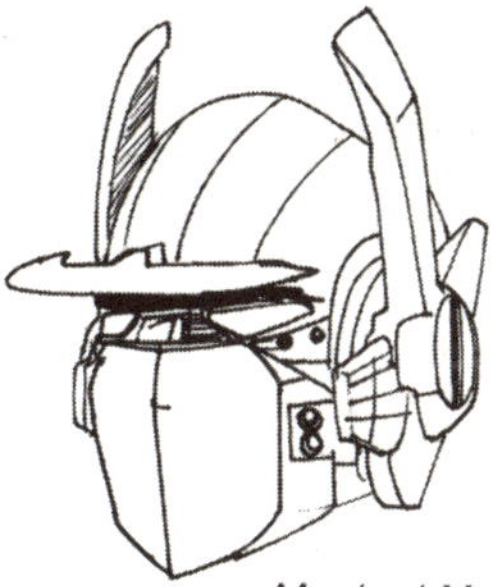

Masked Mouth

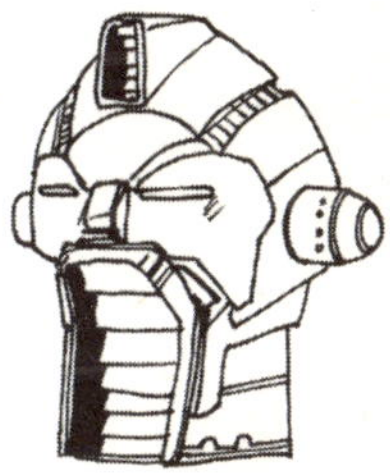

Concave Vent

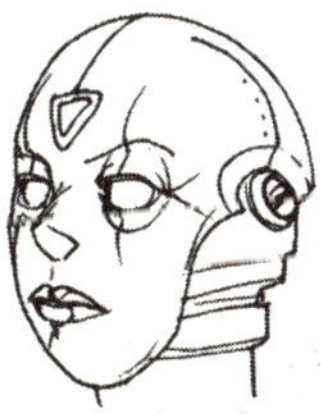

Humanoid-looking Face

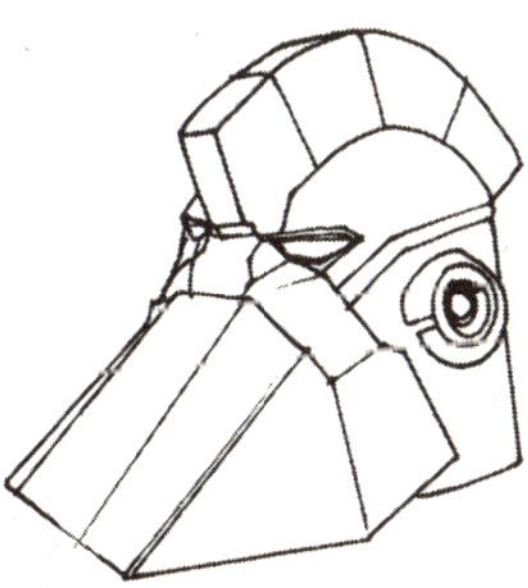

Heavy Protruding Chin

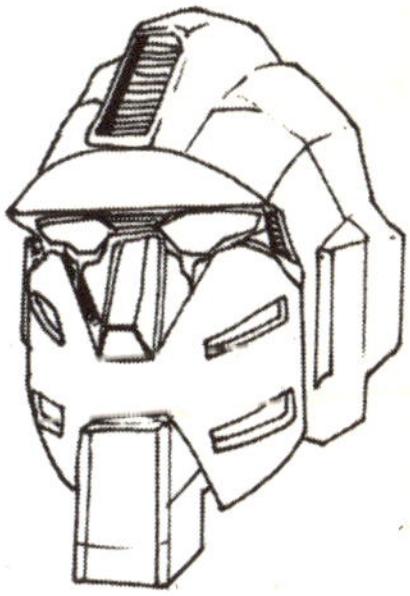

Square Jutting Chin

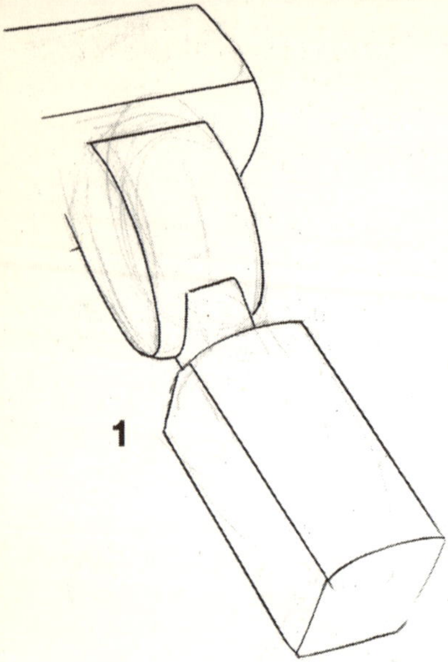

1

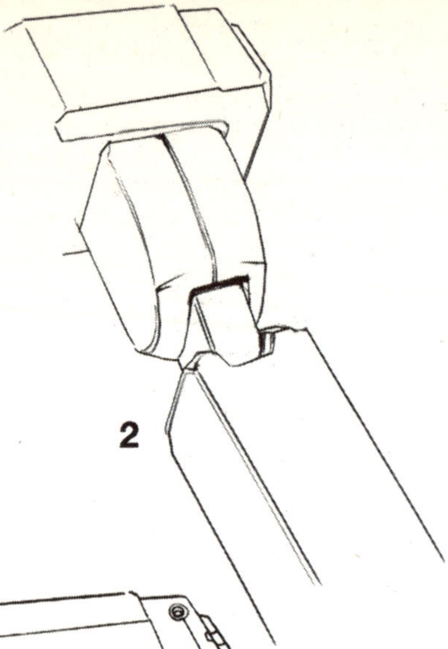

2

The next part of mech design we're going to go over is designing arms. These next three examples are the different stages of increased detail that add weight and mass to a mech. Illustration #1 shows the basic form; there are no plate lines or other details. Notice how the piece looks like it has no weight.

This next example has only a few detail lines added. These added lines on the second arm help the illustration take shape, and the illusion of mass is beginning to present itself.

Finally, cleaned up lines and pieces taken out give the arm its form.

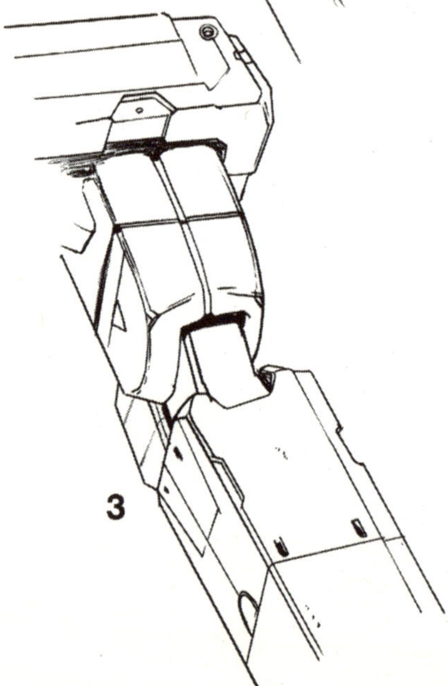

3

GEOMETRIC AND ORGANIC SHAPES

There are two different approaches to designing mech, geometric and organic. Both versions of the arm follow the same anatomical principles. The difference is the geometric arm is more structured and engineered to give it just enough movement to work.

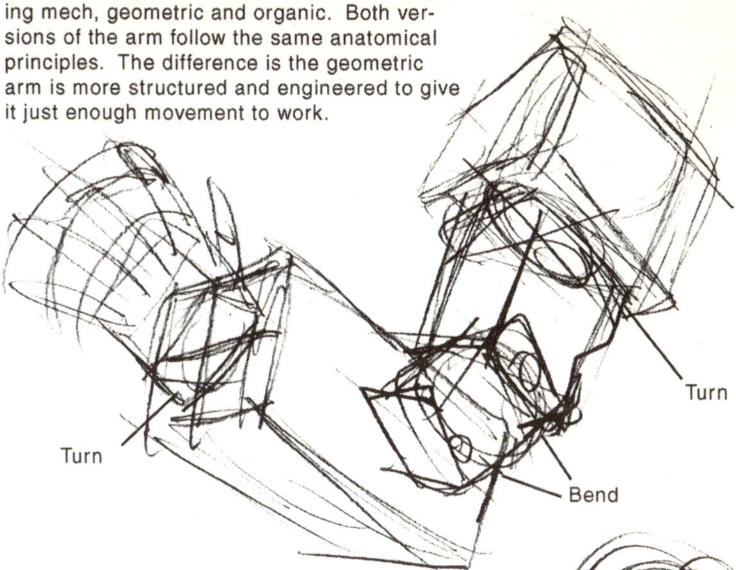

The organic arm is made to be more fluid and dynamic. It performs its tasks more like a human arm.

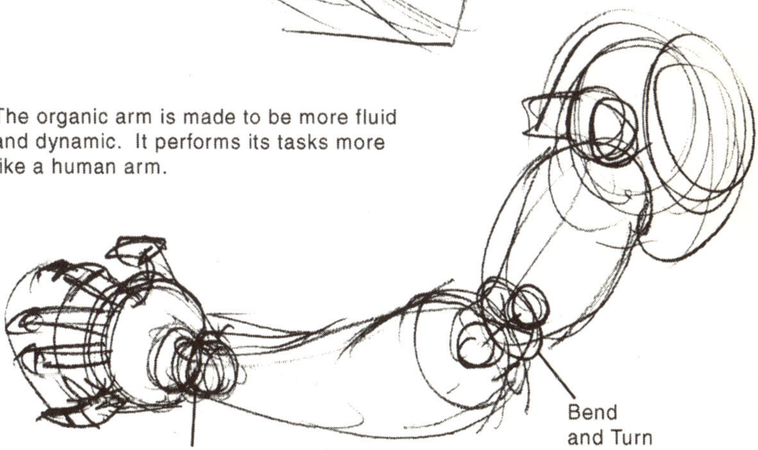

When you get to the rendering stage, you really see the difference between the two approaches of organic and geometric mecha design. In geometric mecha design, 90-degree angles are the main rule. In organic mecha design, angles greater than 90 degrees and curves are used.

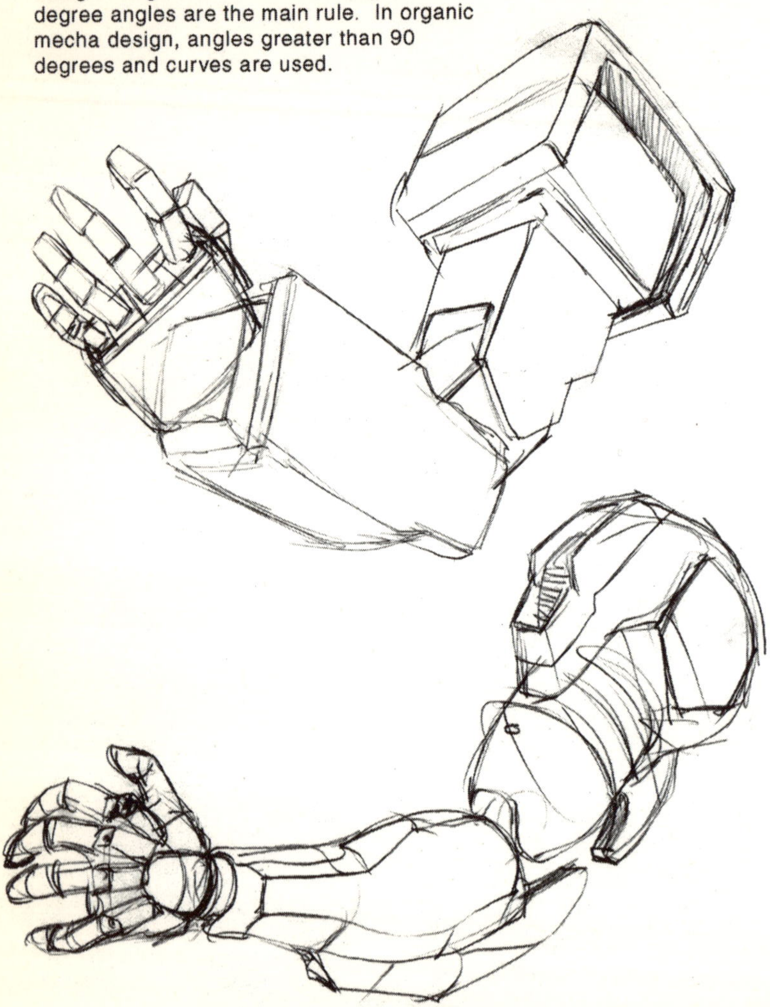

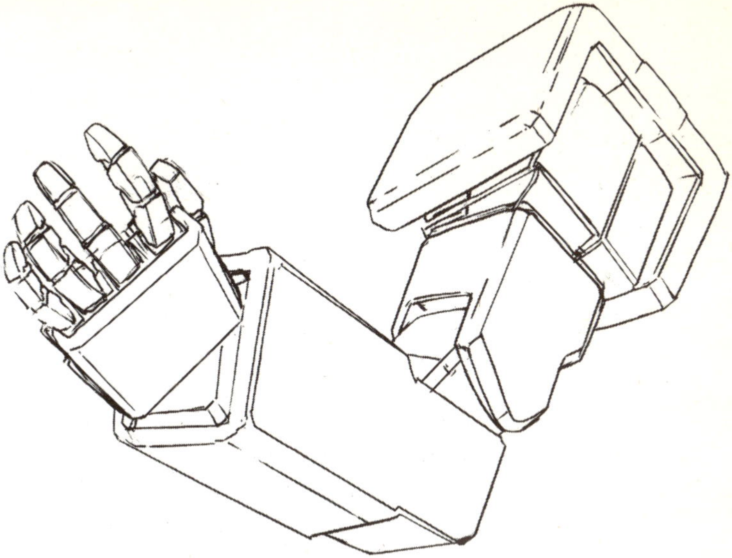

When rendering the drawing, using double lines adds weight to the geometric arm. In the real world, robotics engineers don't just use metal and wires but other materials as well. Think in terms of plastics, composite materials and ceramics.
Plastics and ceramics have a dull finish, so the paneling doesn't need to be so defined.

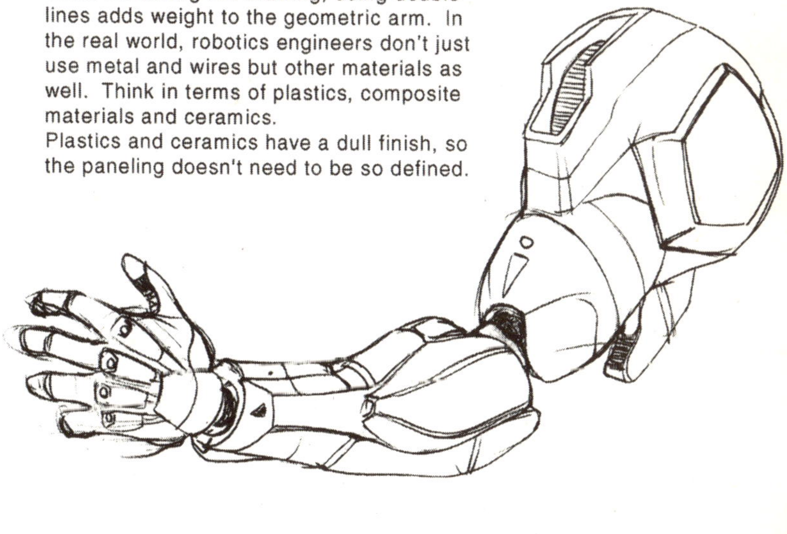

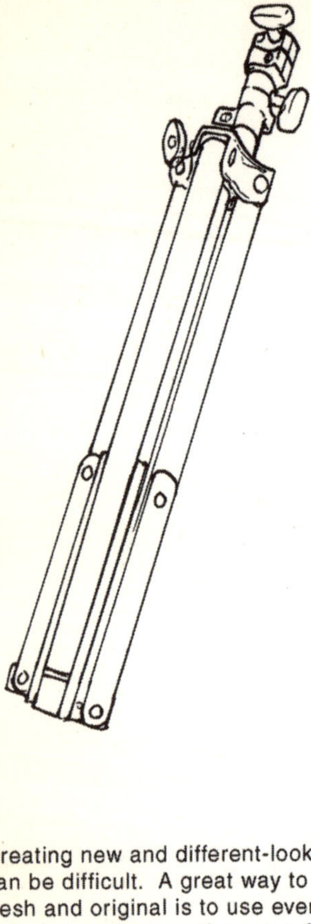

POINTS OF REFERENCE

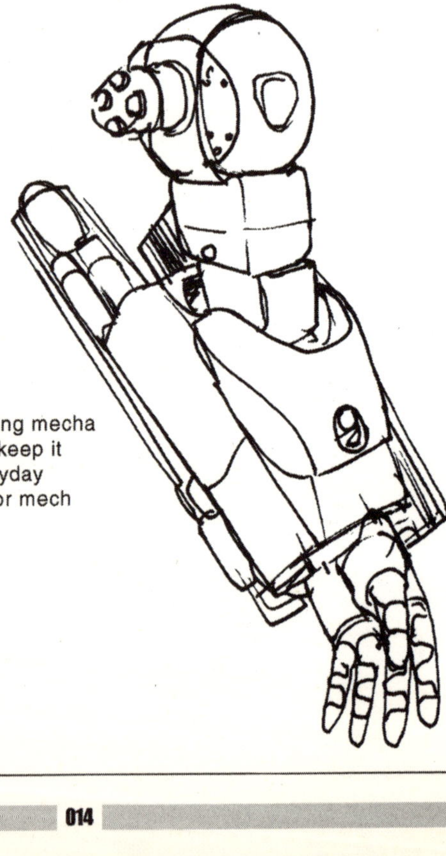

Creating new and different-looking mecha can be difficult. A great way to keep it fresh and original is to use everyday objects to reference the setup for mech part design.

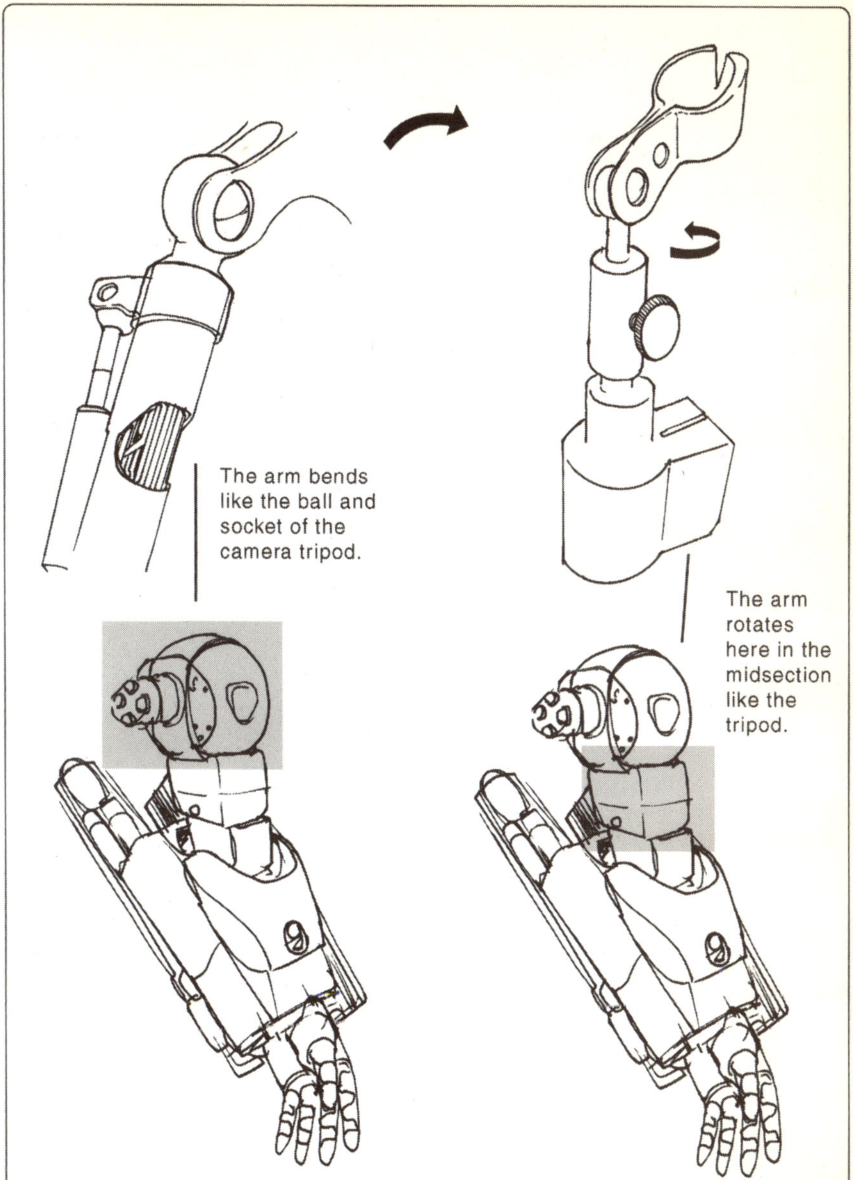
The arm bends
like the ball and
socket of the
camera tripod.
The arm
rotates
here in the
midsection
like the
tripod.

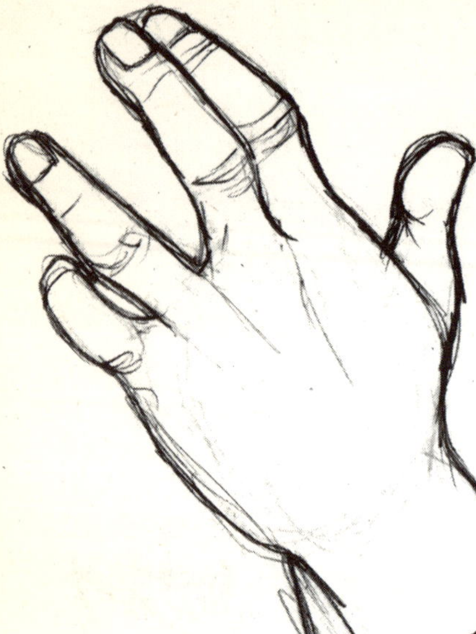

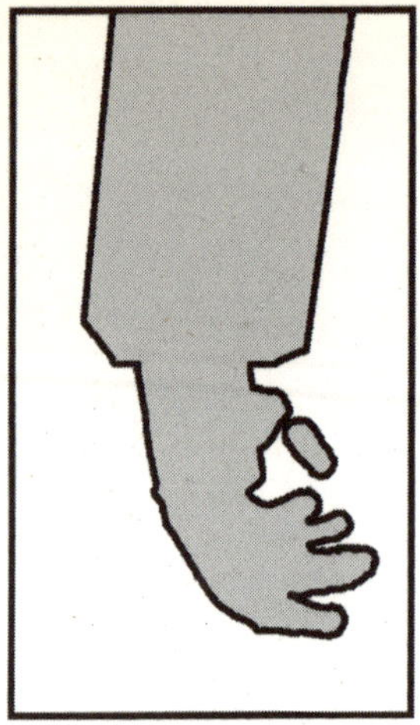

MECHA HANDS

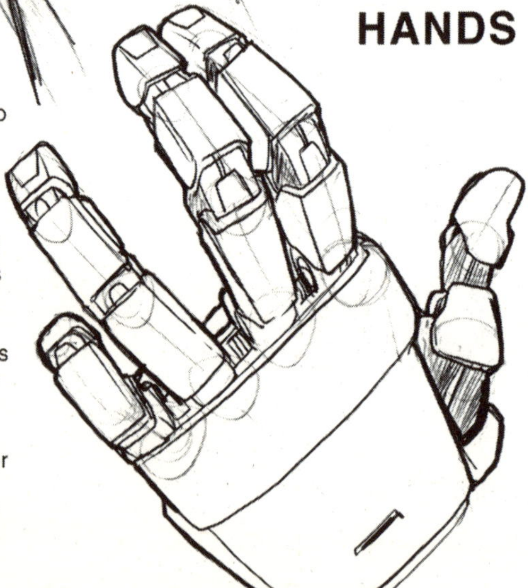

Drawing the hands of a regular person can be tough. Taking a human hand and translating it into a mech hand is even tougher. A good practice to follow is to use your own hands for reference. After all, you've got two of 'em! Picture your hand as a series of long and short cubes, leaving empty spaces where the knuckles and joints would be. After you've finished the first part, you can then fill in the empty spaces by either drawing in gears and motors in those spaces or by faking some fibrous metal material.

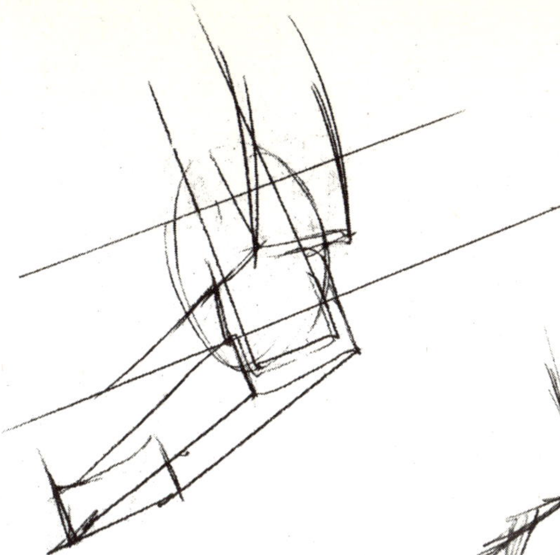

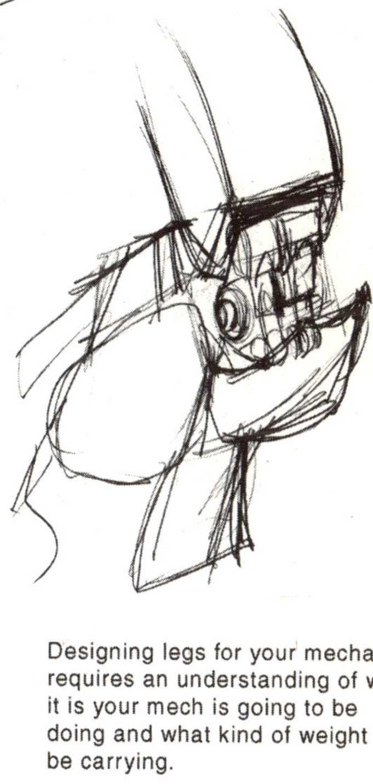

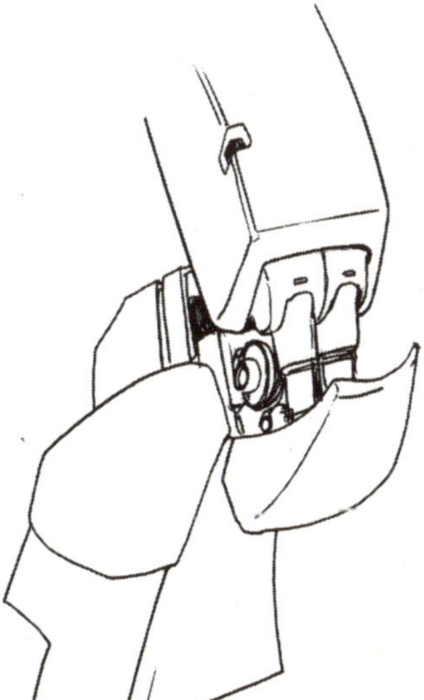

Designing legs for your mecha requires an understanding of what it is your mech is going to be doing and what kind of weight it'll be carrying.

This knee joint is set up like a hydraulic press. Make your design based on what works and, if possible, on things that already exist.

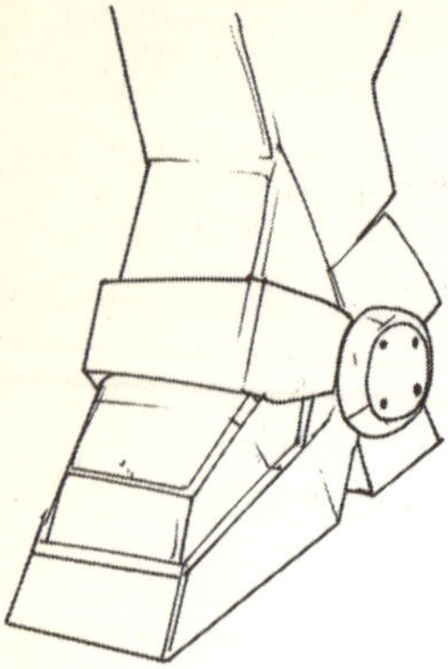

FOOT DESIGNS

The feet help support the mech's body weight.

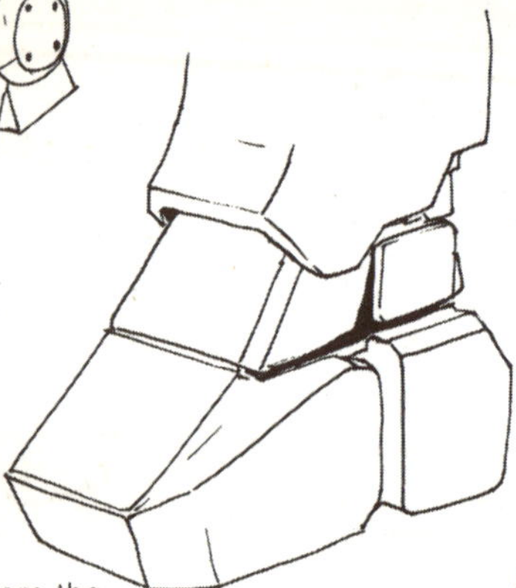

ARCHED

Works well with light, fast mechs

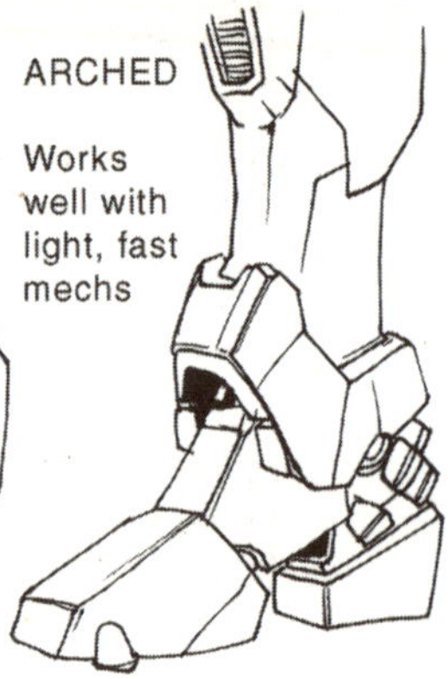

STANDARD

The two feet here are the typical models for a moderately armored mech.

HEAVY

Supports the weight of a heavily armored mech

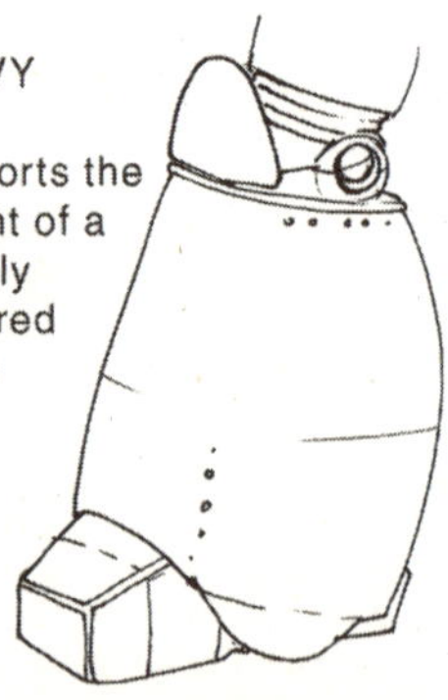

INVERTED

Works well with chicken- or ostrich-style mechs. Also, these legs offer better weight support.

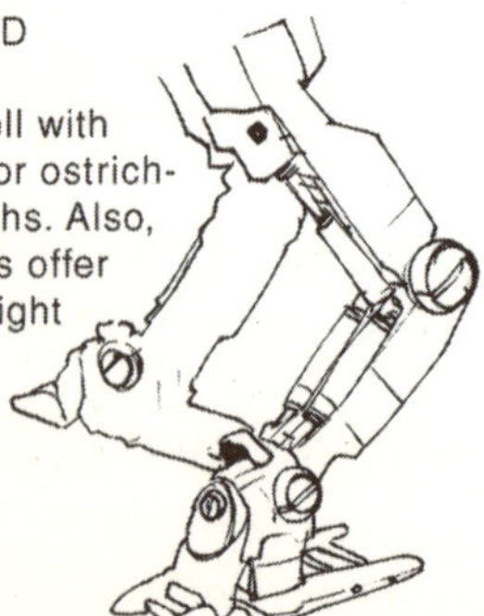

PANELING

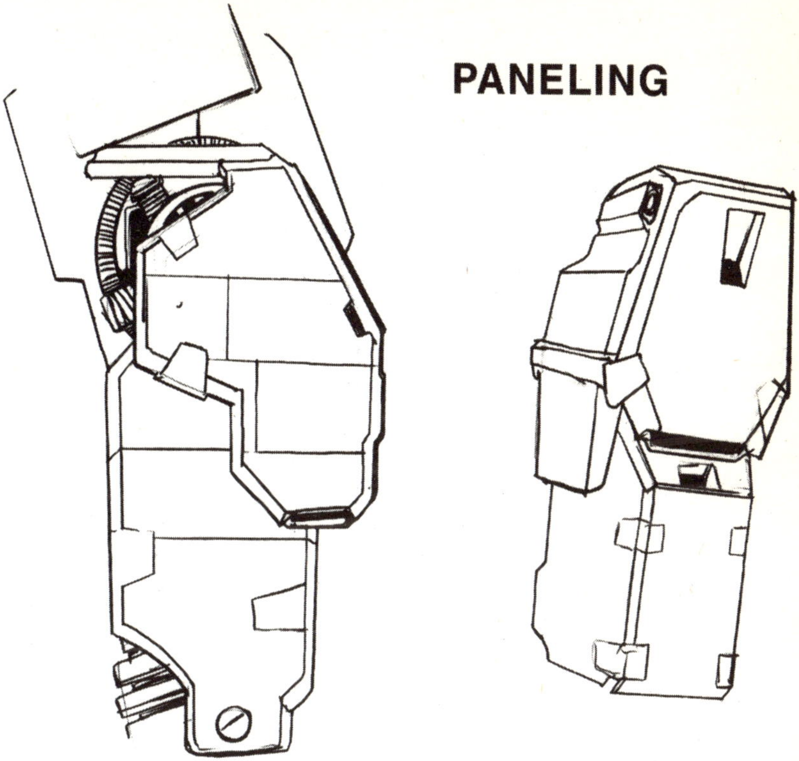

Now that we've gone over a couple of the tougher points of designing mechs, the last one we'll touch on will be the actual plating of the mech. Extra plates are usually designed around the joints as extra protection. Also, you want to consider how the panel's angled, as angled armor prevents deeper penetration by any projectile or shrapnel.

HOW ANGLING MAKES A DIFFERENCE

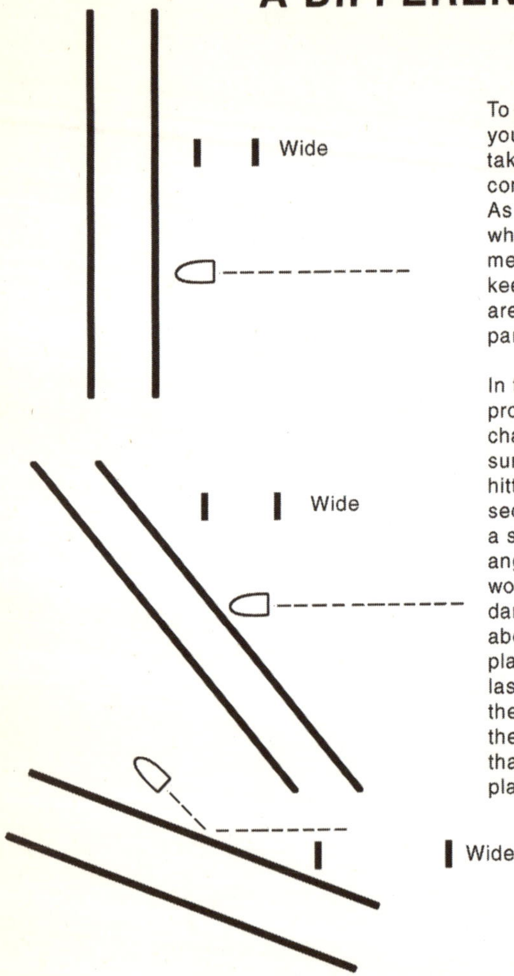

To help with the realism of your project, you have to take a lot of real-world conditions into account. As mentioned before, when you're giving your mech its armor plating, keep in mind that angles are always better compared to flat surfaces.

In the first illustration, a projectile has more of a chance to penetrate the surface because it's hitting head-on. In the second example, there is a slight angle. This slight angle on a breastplate would probably still take damage, though only about half of what a flat plate would take. In the last example, the angle of the plate is so steep that the projectile would more than likely ricochet off the plate.

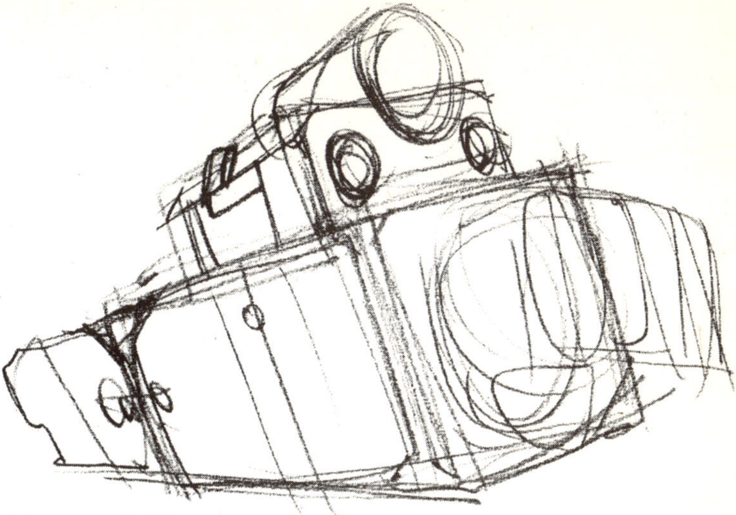

Creating new and different ways to equip your mech can be as simple as attaching the weapons to the arms instead of having them be hand-held. Just be sure when you're deciding on what and where to attach things that everything remains functional and practical.

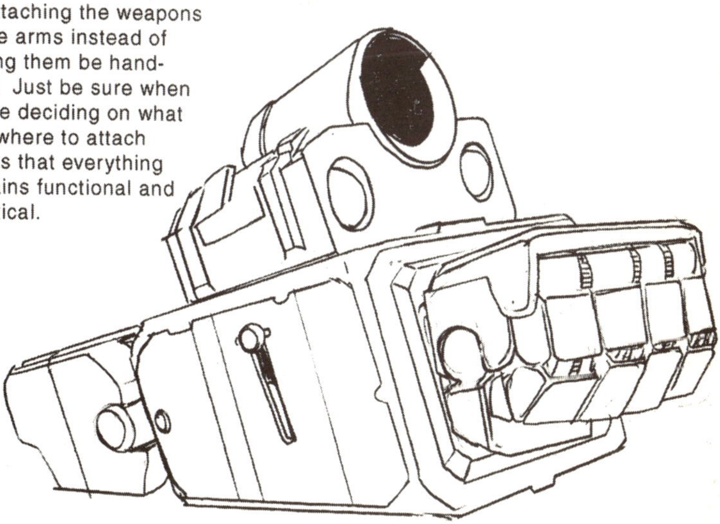

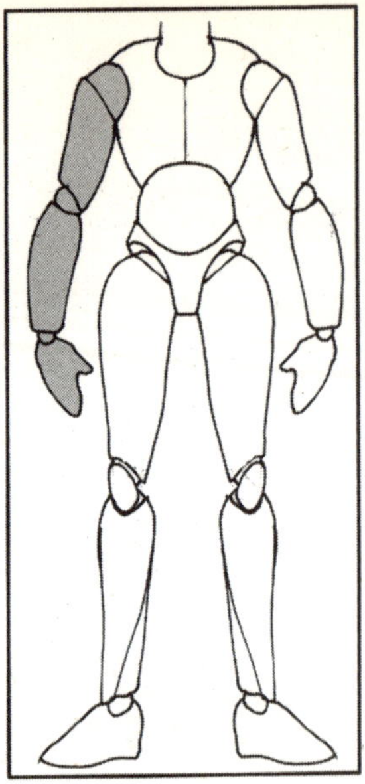

On this figure, the right arm has been replaced with a heavy cannon. A considerable weight difference between the right and left side of the mech's body is to be expected. There should be a counterbalance for this, possibly a shield or a thick left forearm.

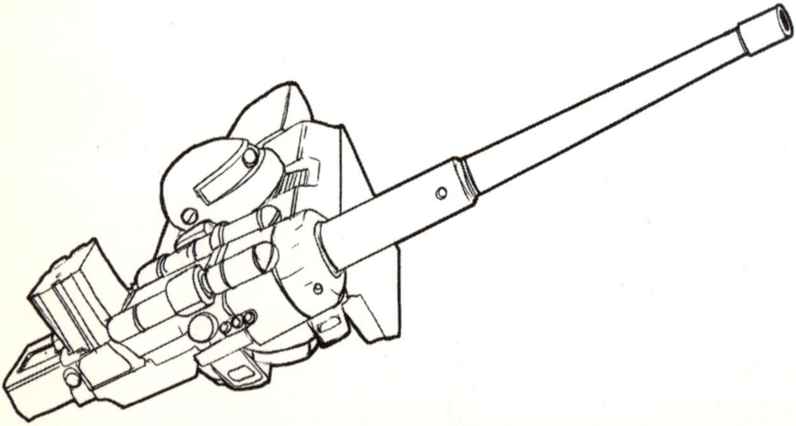

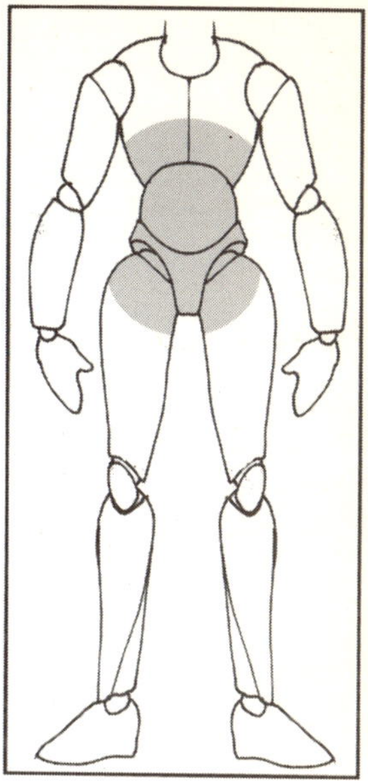

The attachment here is minimal. A booster in the center of the mech's back gives the mech additional speed, but doesn't intrude on any other bodily features.

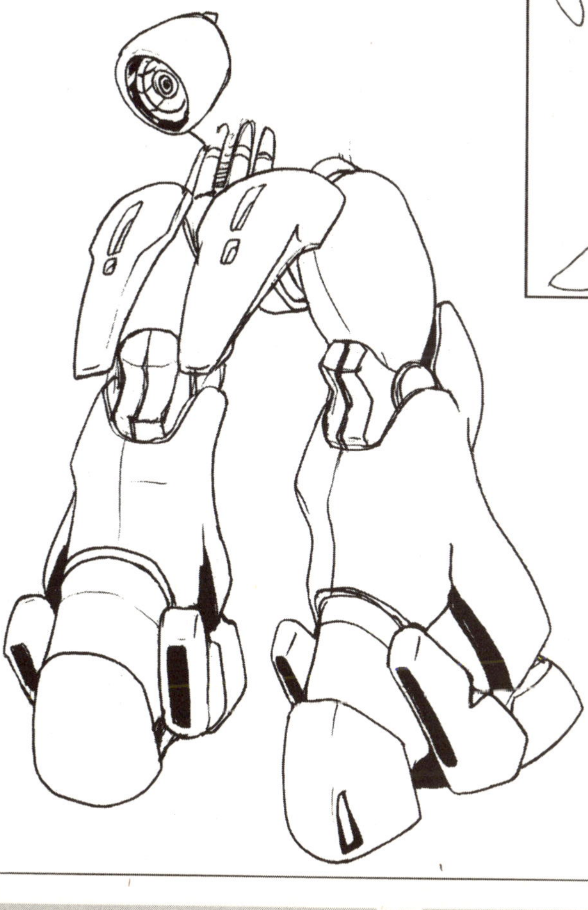

ASSEMBLING PIECES

It will probably be only on rare occasions that you will have to illustrate a mech disassembled and then reassembled. If you do find yourself having to draw this, having a simple understanding of how the different components of a mech go together will prove to be helpful.

The easiest way to tackle this is to treat the mech components with the “tab and slot” approach.

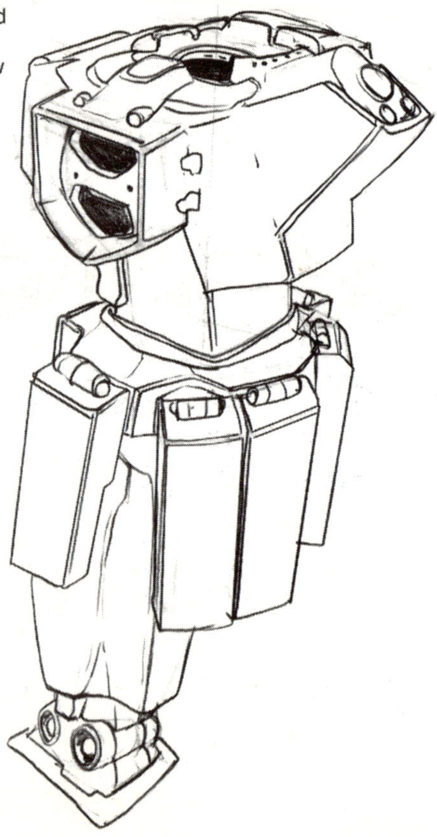

Notice the siots in the torso where the shoulder will be. On the arm, there are corresponding tabs that are drawn to fit into those slots. Even though your viewers will probably never see those tabs and slots, it is important for you as an artist to know that they are there.

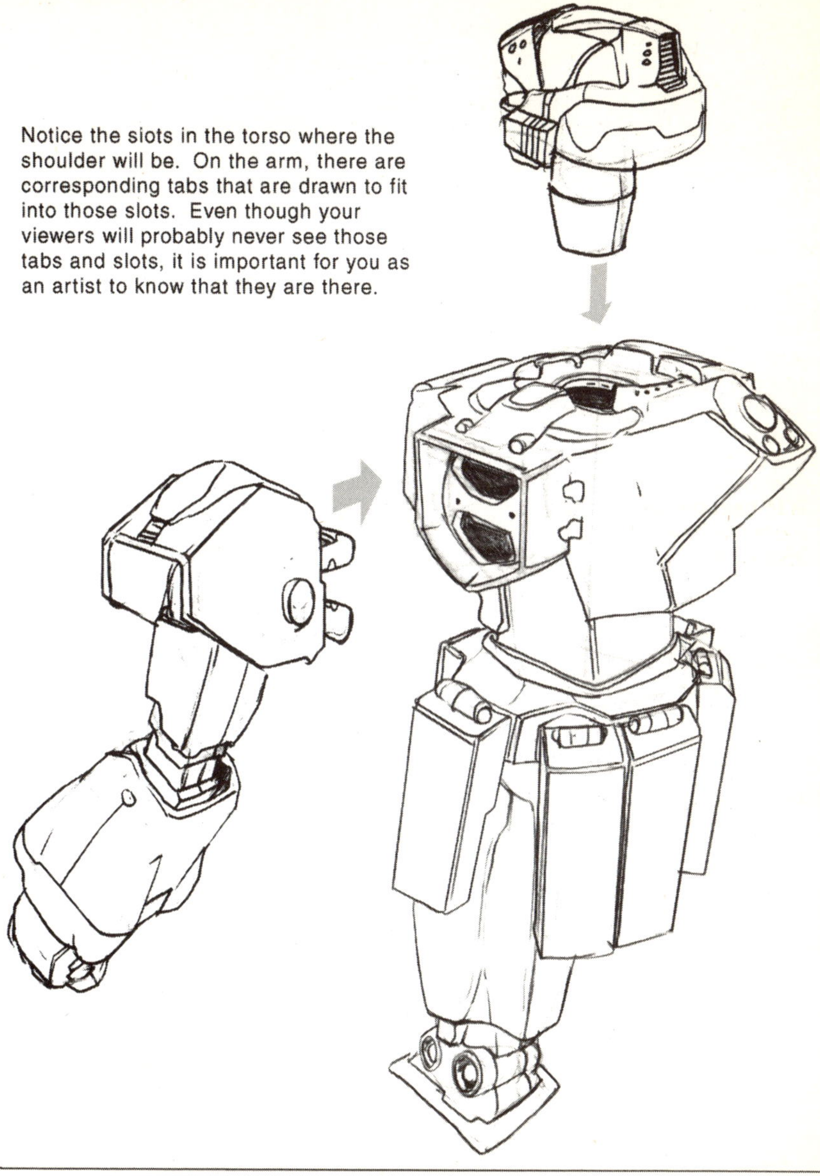

Your mech should look solid after the components are attached. The design of your mech should suggest to your viewer where the tabs and slots are without ever showing them. This will aid you later when you're illustrating a repair or damage happening to your mech.

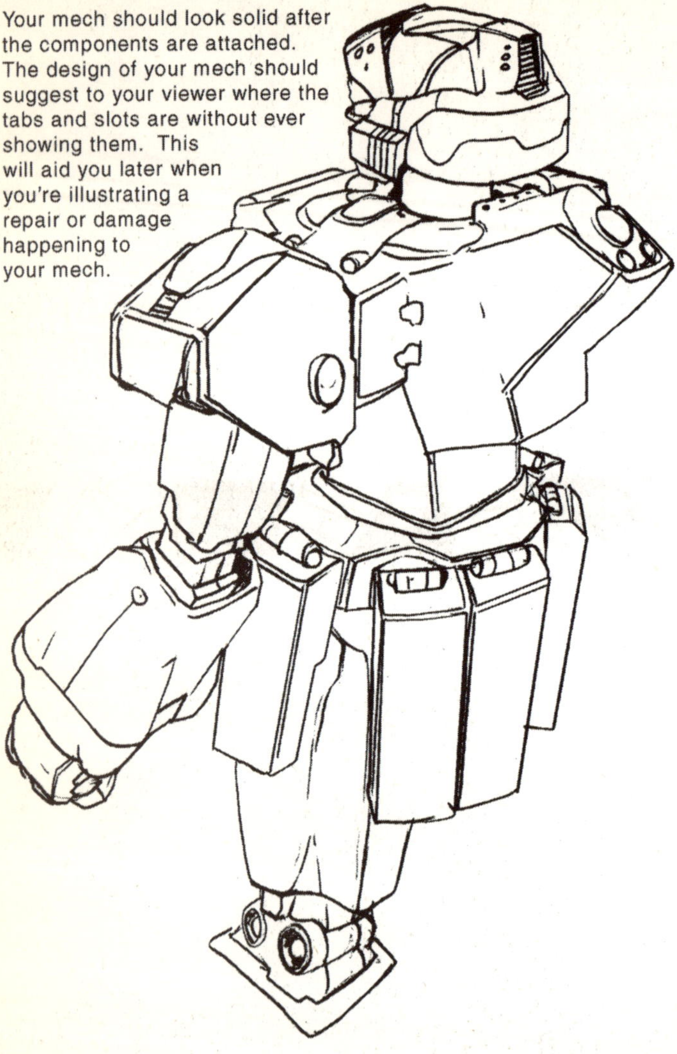

CHAPTER 02

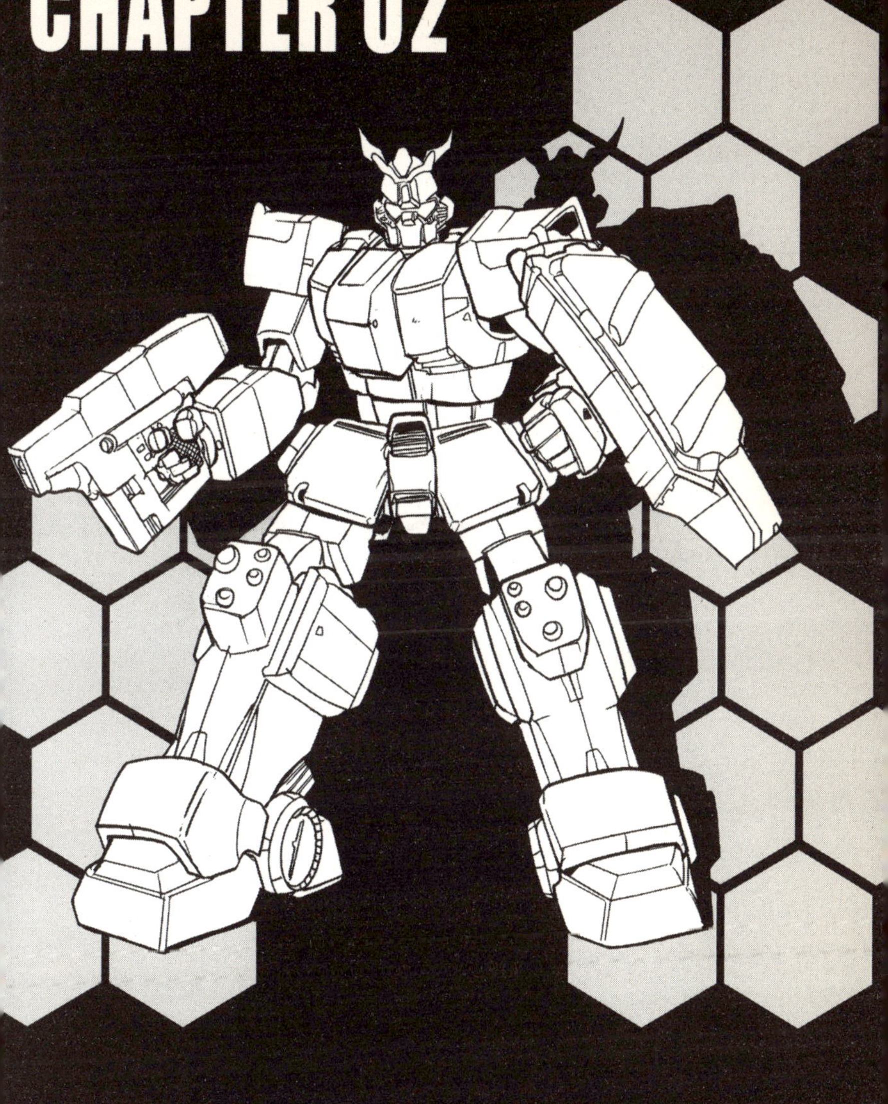

DESIGNING FACTIONS

HEROES VS. VILLAINS

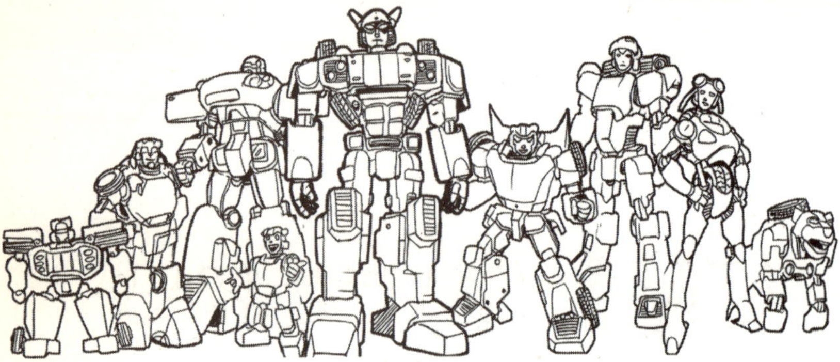

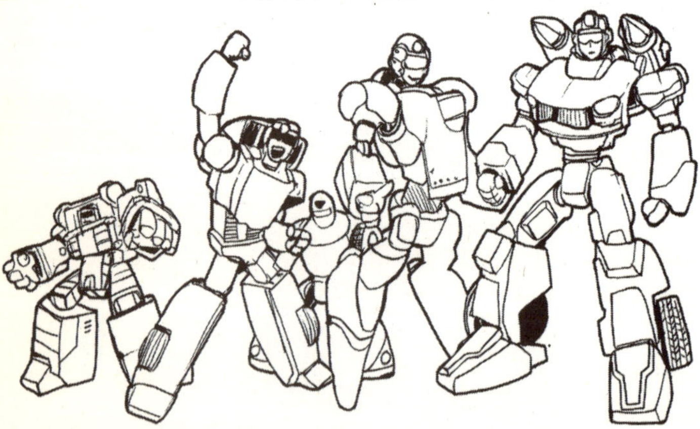

When creating a faction or team, first you've got to decide whether you're designing heroes or villains. When you're designing, the heroes are usually larger in number.

In comparison, the villains are fewer in number, but their size and strength definitely make up for the difference.

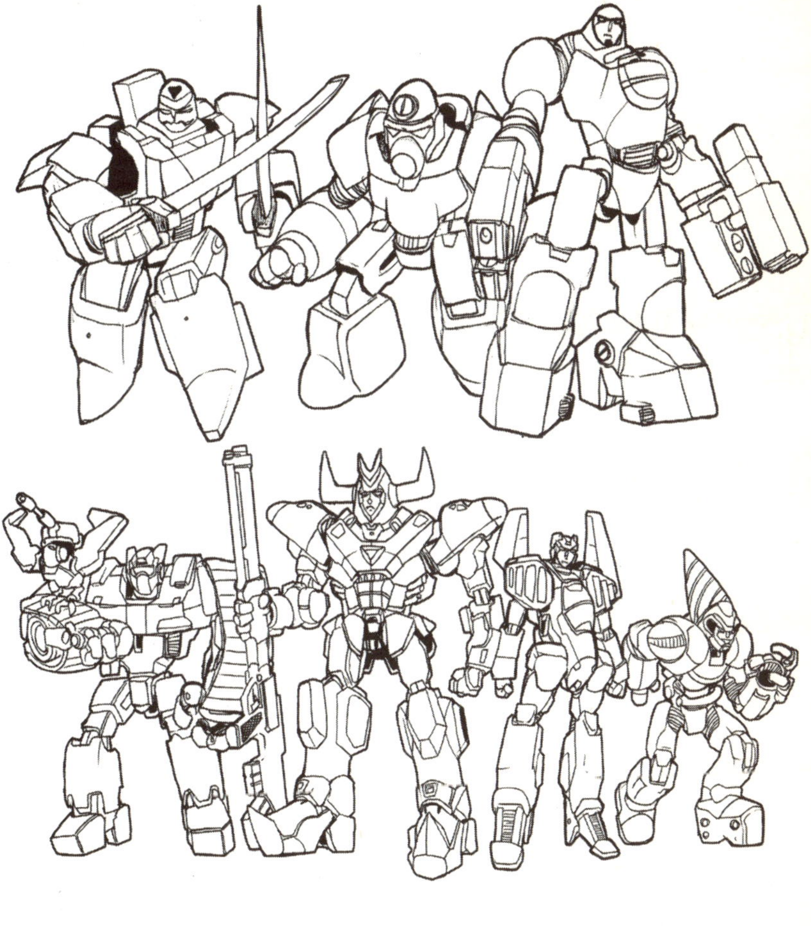

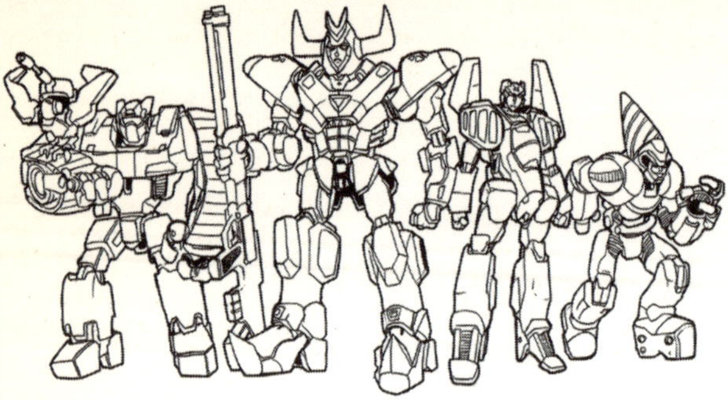

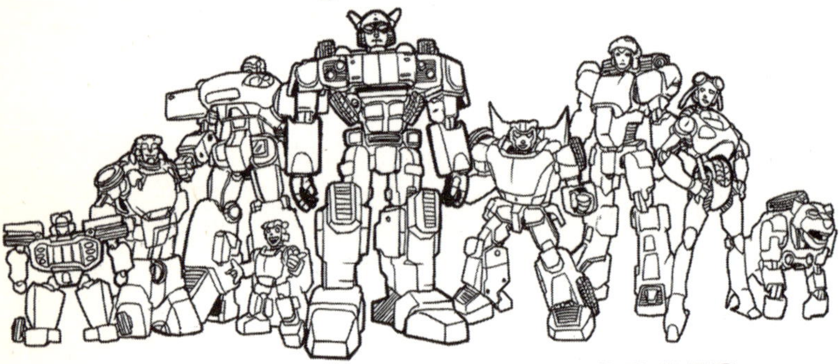

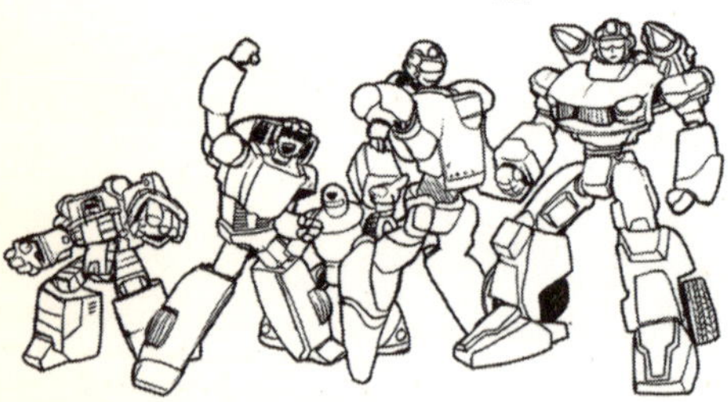

ROBOTS

VS.

ALIEN ENEMY

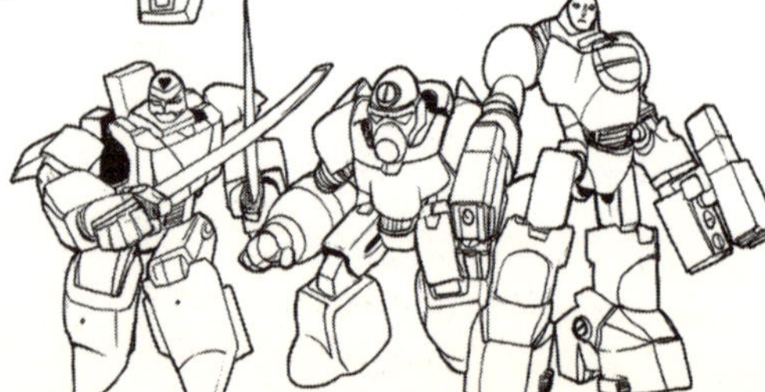

Besides the usual robot-versus-robot scenario, there's the good and evil team-up against a completely new menace.

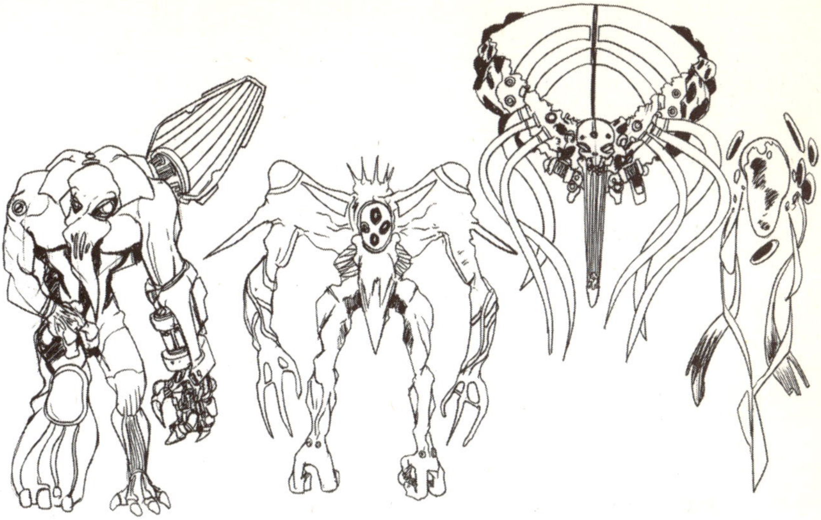

The new threat takes on an entirely unfamiliar shape, so the differences in the designs of the good and evil robots on the combined team aren't that striking.

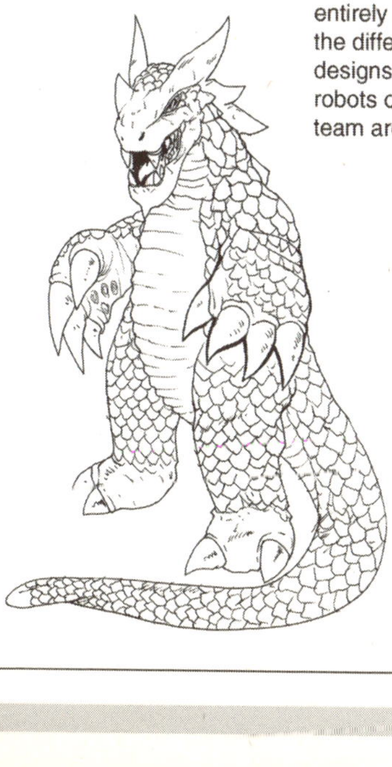

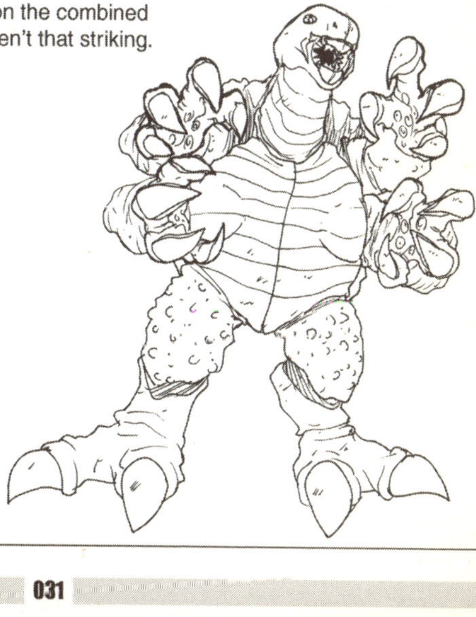

Yet another spin on the heroes-versus-villains theme is the situation in which the heroes face an indeterminate number of enemy robots. In this case, it could almost be assumed in your storyline that there is a place where the enemy is continuously mass produced. So, the main plot of the story is for the heroes to locate the factory and destroy it before the enemy takes over. The great thing about this type of scenario is that your designing duties can be more focused on the personalities of the heroic mechs.

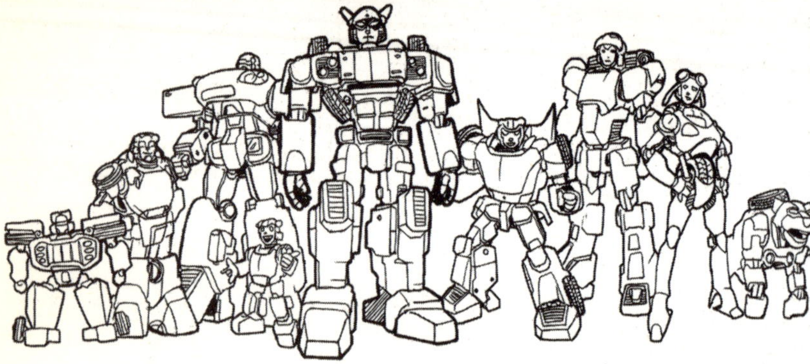

HEROES **VS.** **VILLAINS**

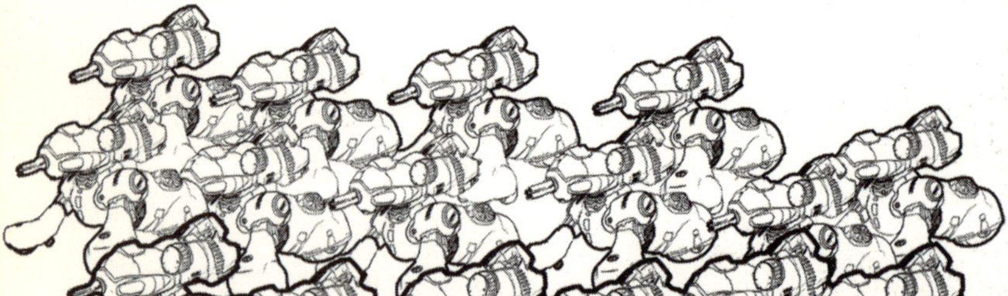

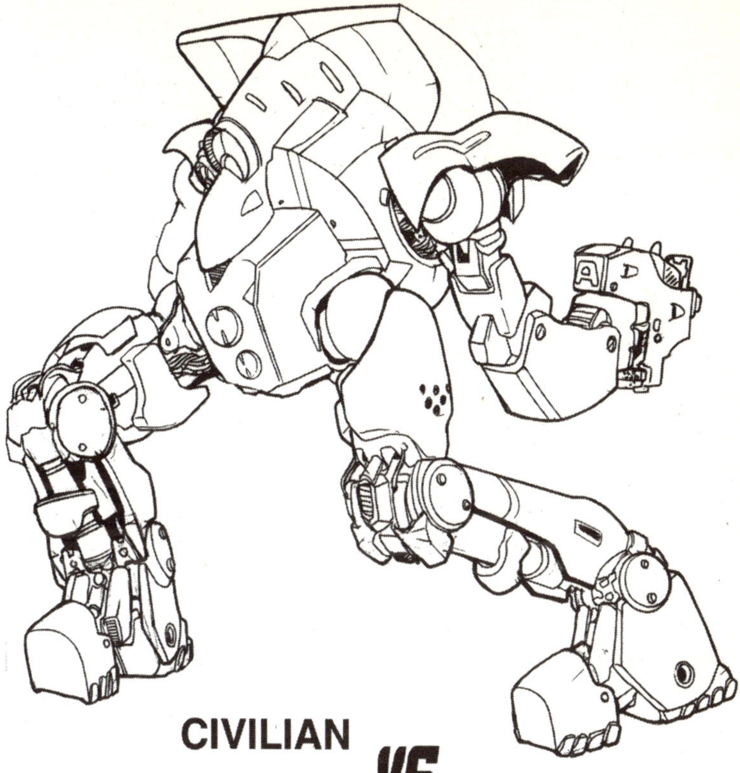

CIVILIAN *VS.* MILITARY

Something to keep in mind when you're designing mecha is the original use for the mech. In the case of a civilian mech as opposed to a military mech, the civilian mech is built for a specific purpose, like moving freight, construction, or in some cases, police duties.

The main difference between the two is that the civilian mechs will usually look more nimble because of the obstacles and humans they have to work around. In most stories in which these types of mech are used in combat, they're jerry-rigged for it.

The military mech is a fighting machine more like a tank, built specifically for the purpose of combat and nothing else. Your design should reflect the fact that this machine is expected to get hit and shot up, so none of the delicate inner workings are exposed, unlike its civilian counterpart. It has to look functional as well; every component has a purpose or else it isn't there.

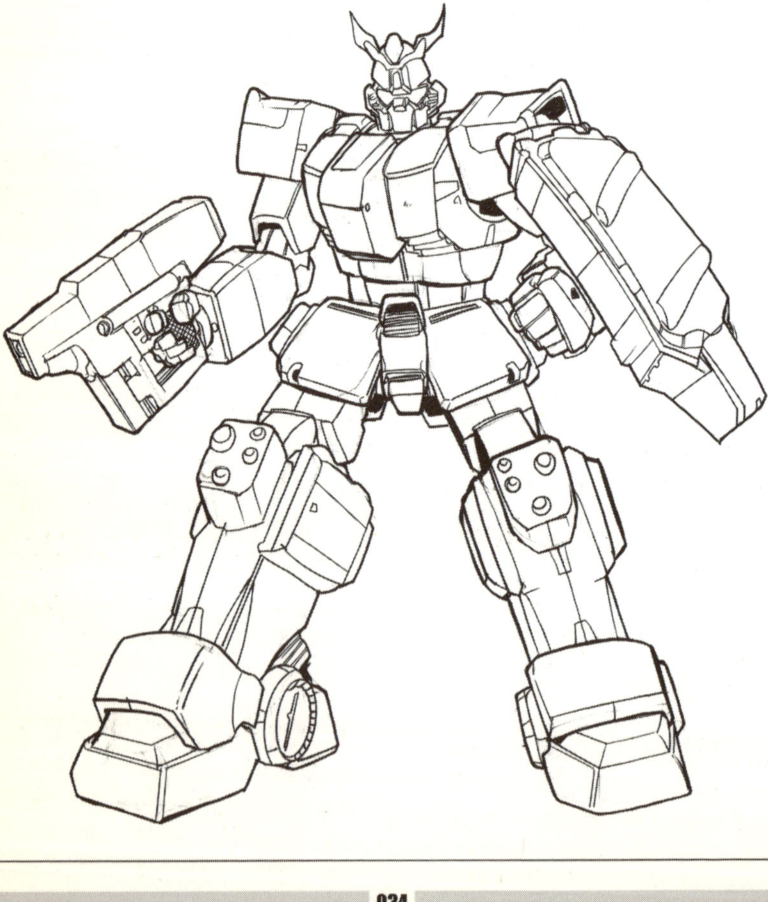

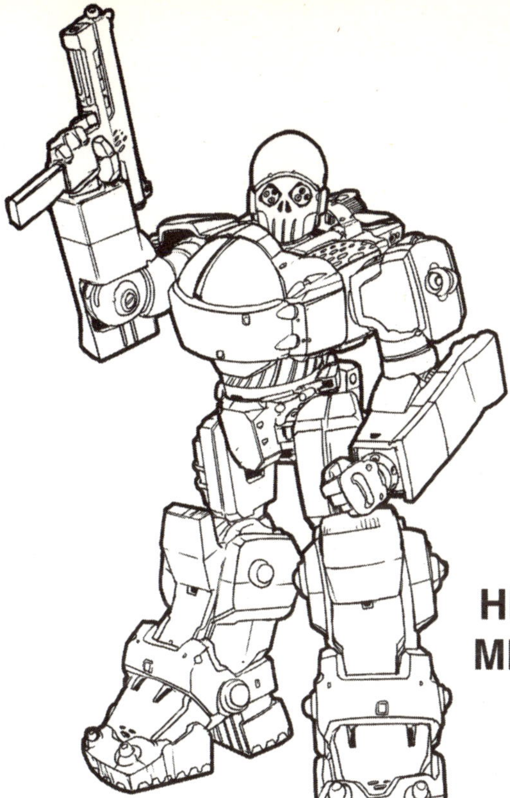

Here, we can use the old ant-versus-grasshopper analogy, except in this case, the grasshopper happens to have a human operator and weighs about 60 tons. Pound for pound, this heavy mech could take on the smaller attack drones one on one and win. They're smaller, and have thinner armor and don't possess the intuition the piloted mech has. But in large numbers, the unmanned drones would eventually deal out enough damage to put this big guy on the repair rack for a while.

HEAVY MECHA

VS.

MULTIPLE ATTACK DRONES

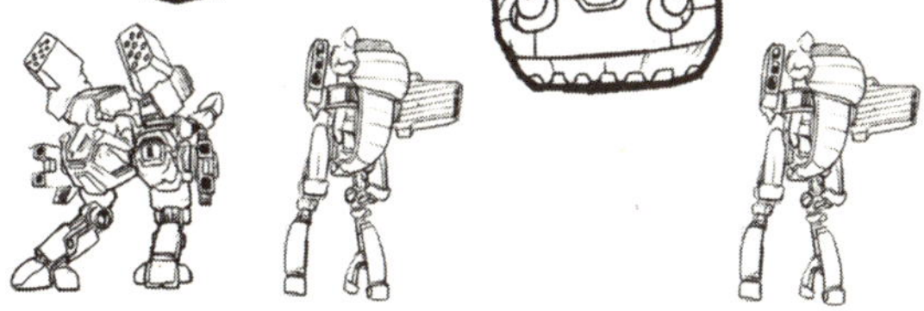

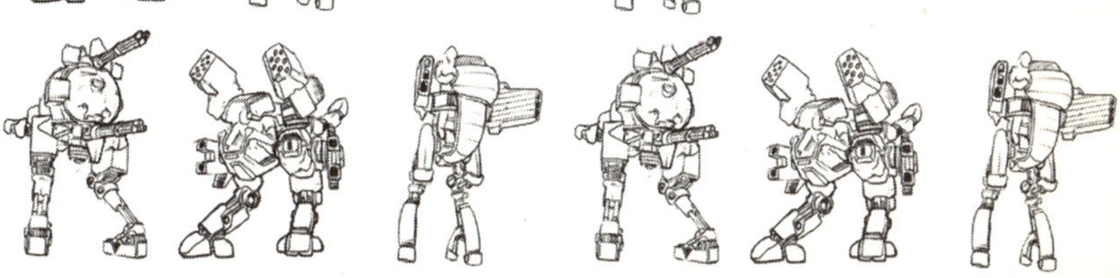

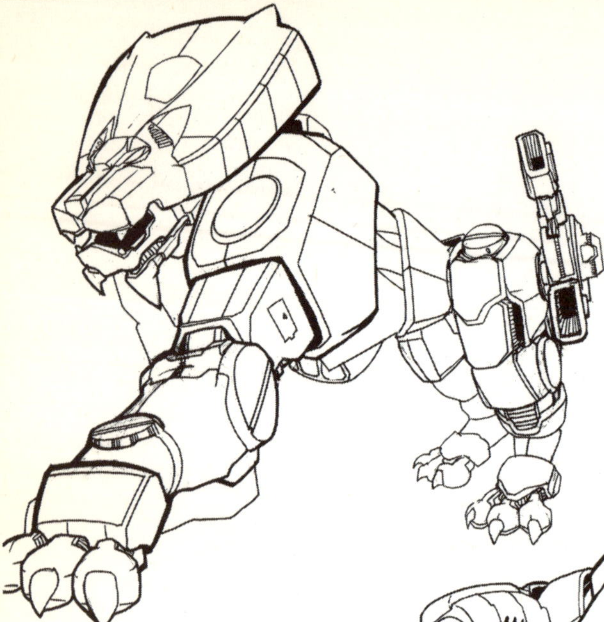

Predator versus prey is a common theme. However, the 'prey' here is fit enough to handle the predator. The predator's armament can acquire an attack posture with ballistic weaponry, while the prey has a defensive posture with edged forearm weapons.

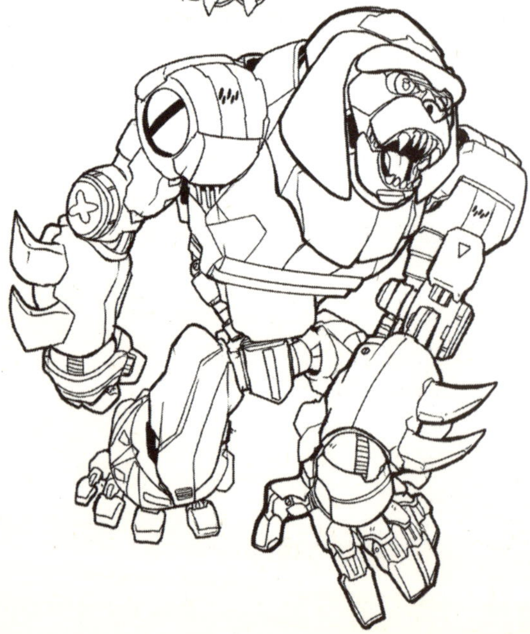

PREDATOR

VS.

PREY

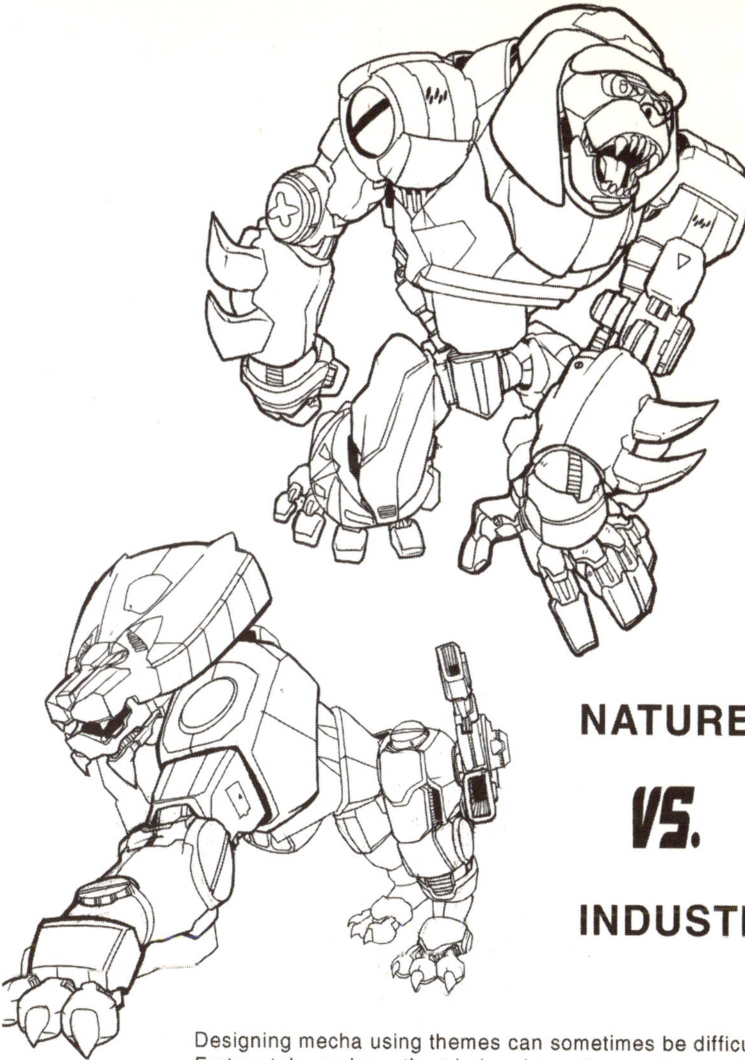

NATURE

VS.

INDUSTRY

Designing mecha using themes can sometimes be difficult. Fortunately, we have the tried and true theme of nature versus industry.

Mechs that are based on the anatomy of animals are the design polar opposite of this tank, so automatically, there is a visual conflict.

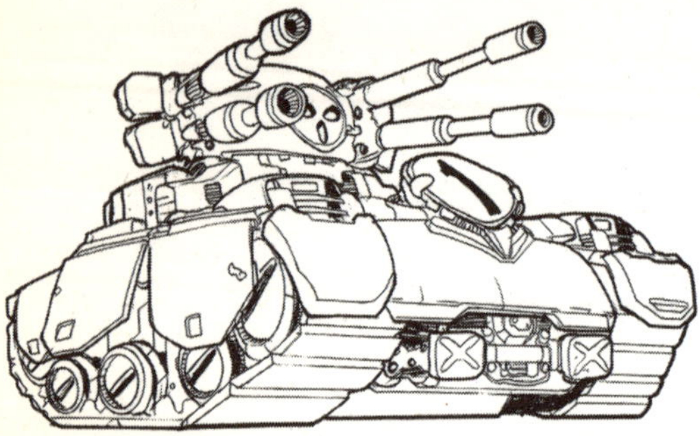

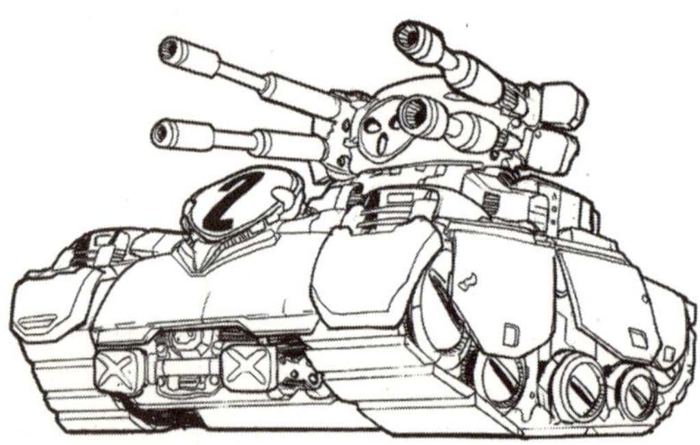

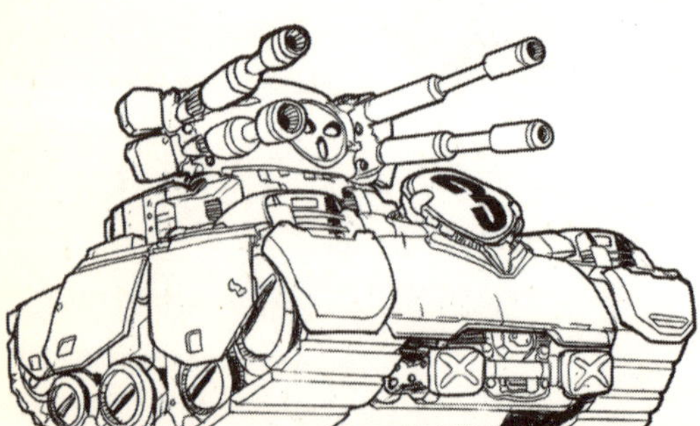

SUBTLE DIFFERENCES

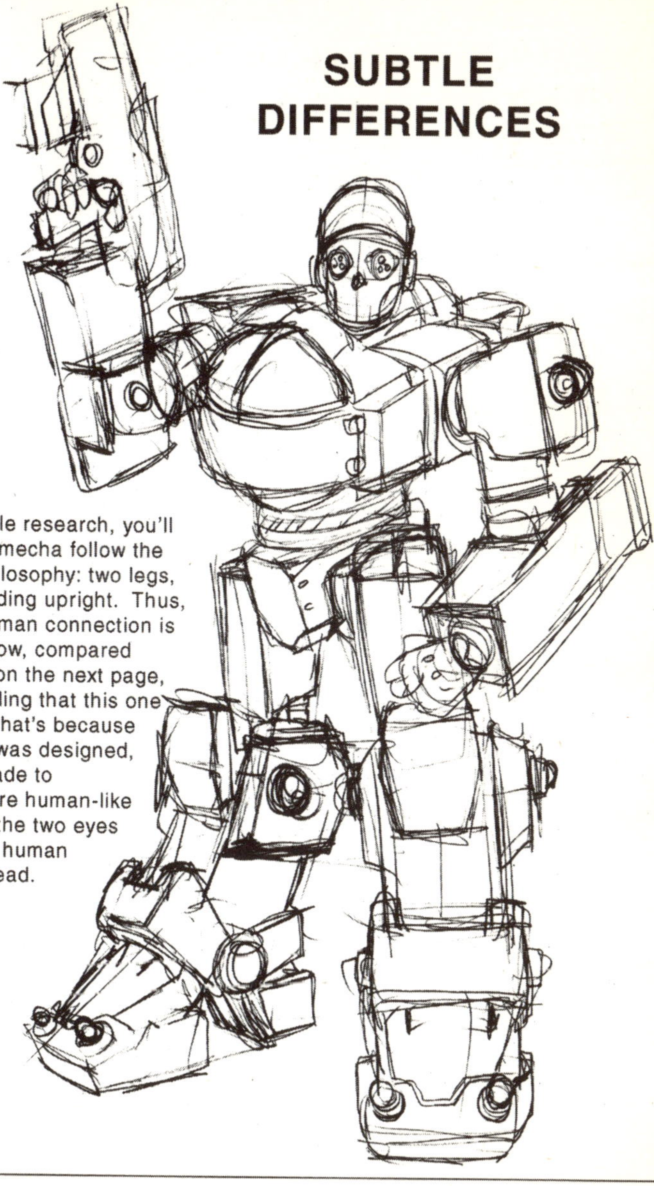

As you do a little research, you'll find that many mecha follow the basic biped philosophy: two legs, two arms, standing upright. Thus, making that human connection is easy. Notice how, compared with the mech on the next page, you get the feeling that this one is the "hero." That's because when this one was designed, efforts were made to incorporate more human-like traits, such as the two eyes and the almost human skull-shaped head.

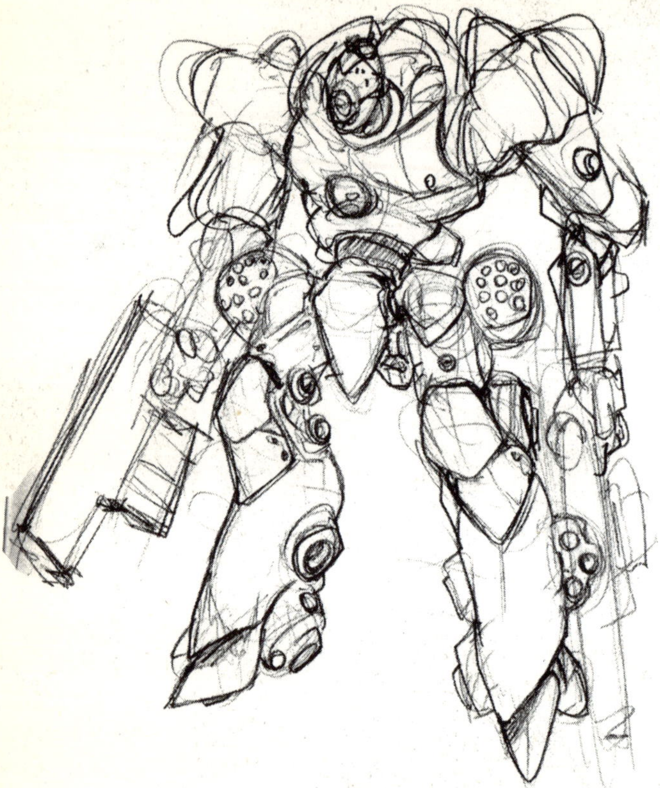

The body styles of this mech and the last one are pretty similar, as they follow the same rules of basic human anatomy. The biggest difference is that the head and face here were deliberately made to look alien. Simply designing a head without the typical human features like the nose or mouth and giving it only one "eye" makes the mech very alien-looking. In some cases, designing your mecha like this can help you achieve an alien or futuristic look.

CHAPTER 03

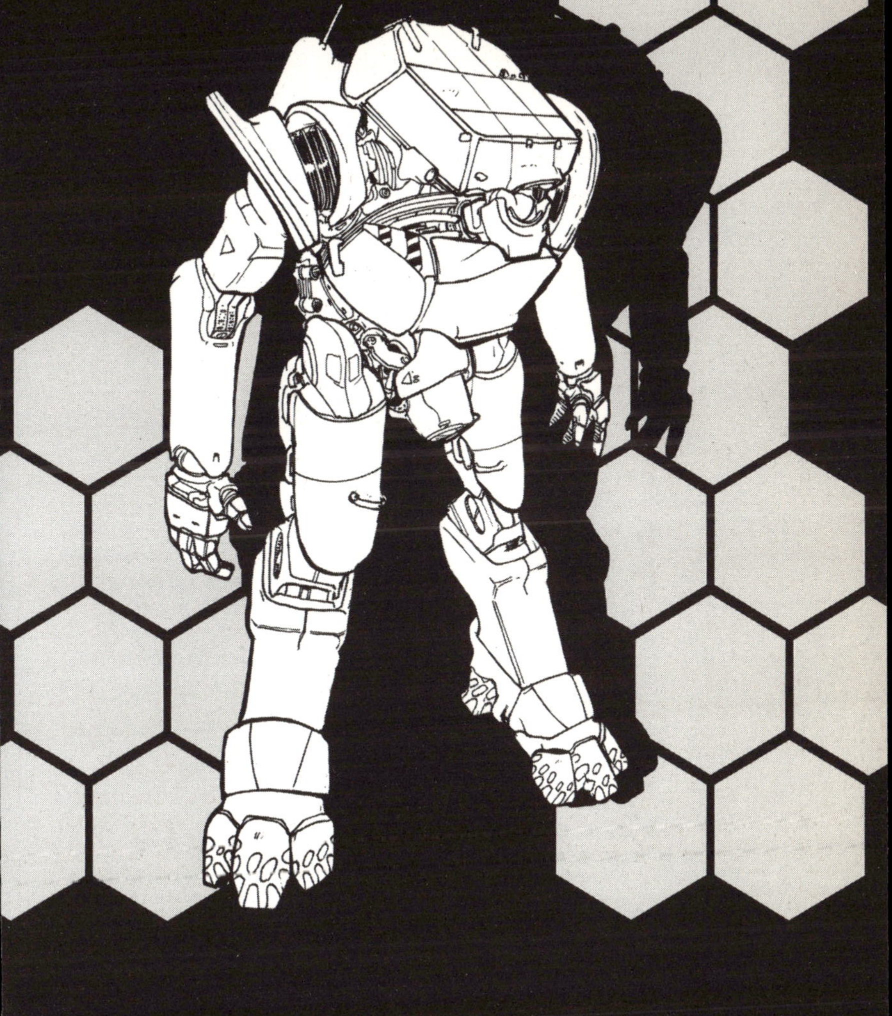

SIZE RELATIONS

Determining the size of a mech is very important. Whatever size you make your mech will determine how heavy it is as well as how practical it would be on the battlefield.

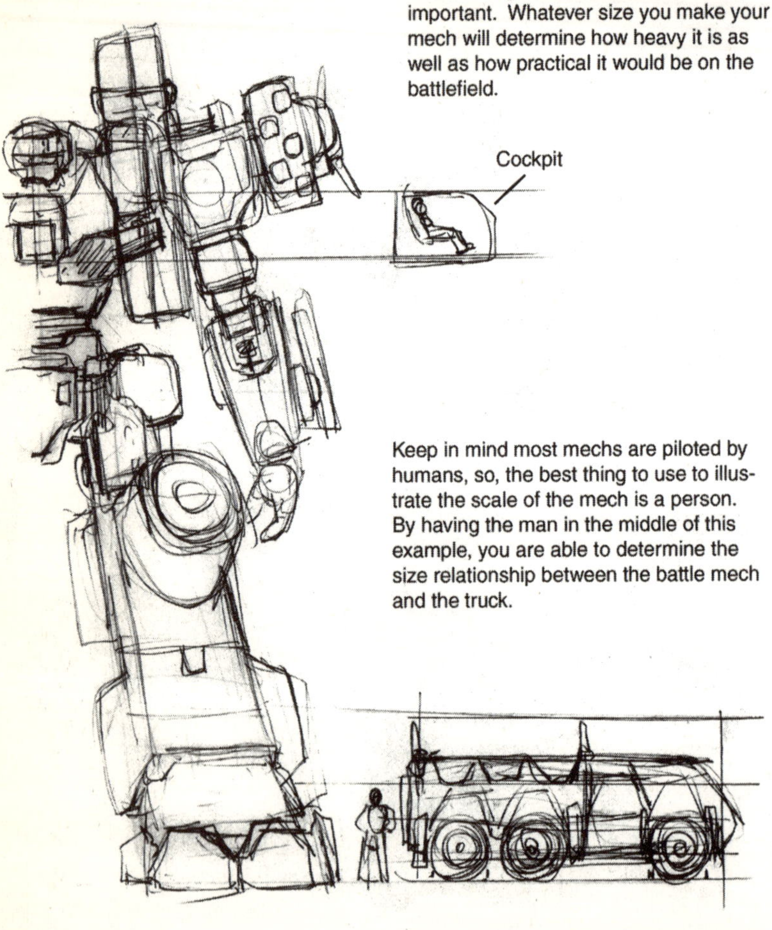

Keep in mind most mechs are piloted by humans, so, the best thing to use to illustrate the scale of the mech is a person. By having the man in the middle of this example, you are able to determine the size relationship between the battle mech and the truck.

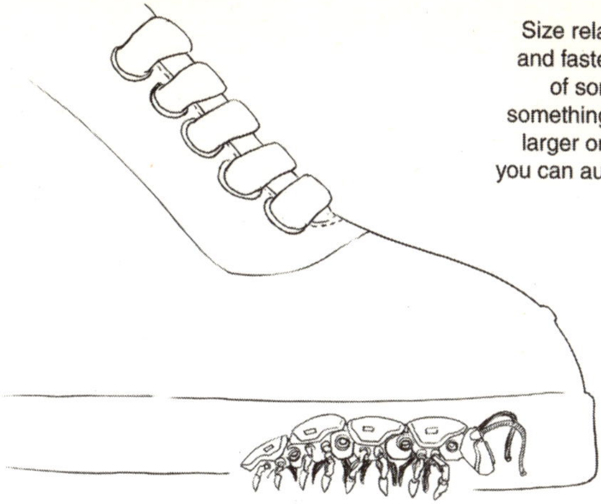

Size relationships are the easiest and fastest way to show the scale of something. When you draw something easily recognizable and larger or smaller than your mech, you can automatically communicate the mech's size.

In the first example, a person's shoe is used and drawn very large in relation to the mech. In contrast, the second example shows an entire person in relation to just a leg of the mech. Which one looks bigger to you?

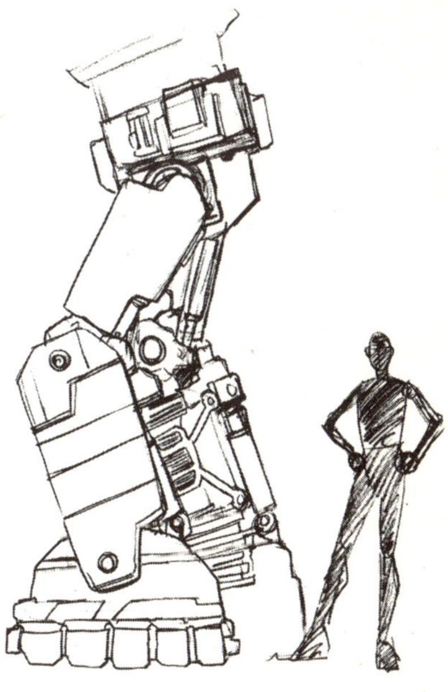

When determining what function a mech serves, we see these tiny mecha could possibly be used as infiltration units or repair robots.

BODY REPLACEMENT

When it come to form & function, a good rule of thumb is to KEEP IT SIMPLE! Keeping it simple when designing is as easy as asking yourself a couple of questions. First, what's the purpose of the mech you're going to design? Something for transport, for battle, for sightseeing? Once you've decided that, then ask what attributes are common to things used for that purpose.

When looking at these two designs, can you tell which one is the transport and which one is used for combat? What made you come to those conclusions?

The armored personnel carrier on the left is designed for speed and cargo, human or otherwise. It's built low to the ground with a wide center of gravity to prevent tipping. The combat mech here is obviously designed to pick a fight. With dual heavy guns instead of hands, a warm handshake isn't in its programming.

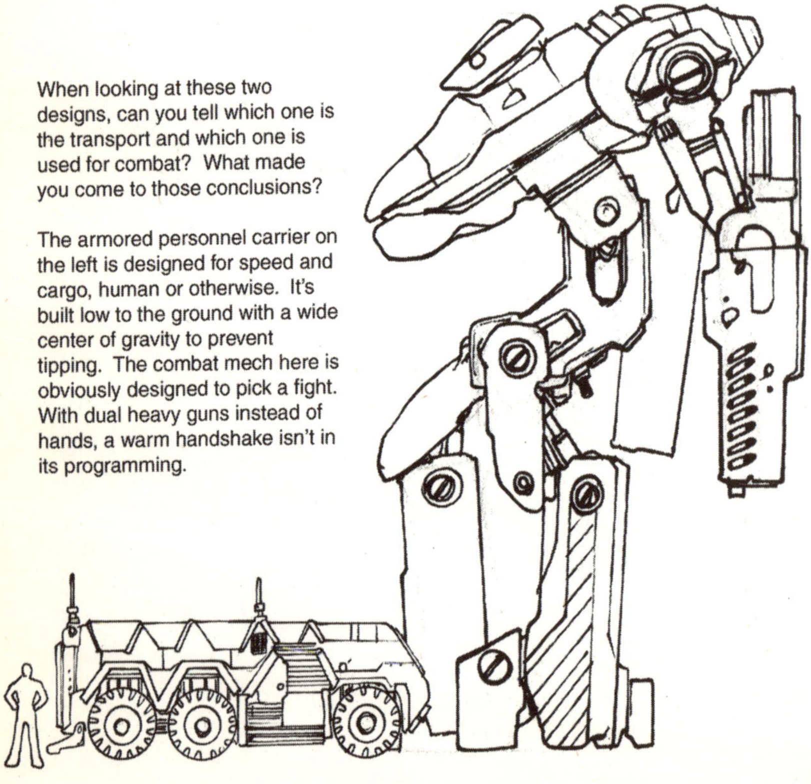

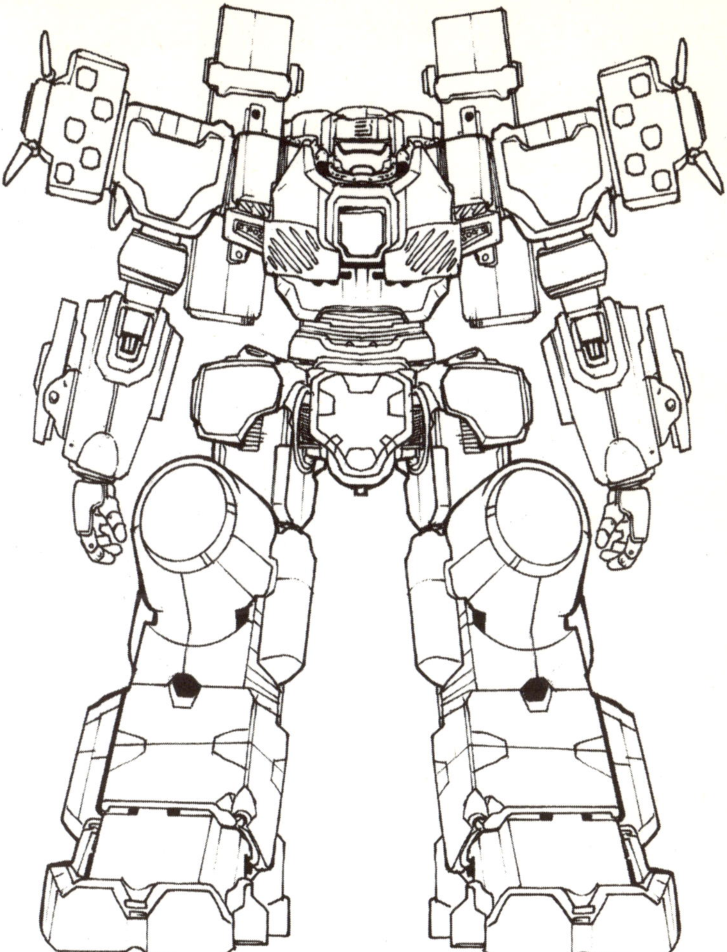

The sheer mass of this mecha is enough to illustrate how heavy it could be. If it's a city-crusher, you want the design to show that it could carry enough ordnance to destroy a city. The one thing to keep in mind when designing is that the perceived weight of the mecha should be taken into account in the structural design. The mid-section and legs should look like they *could* support all that weight. This will help make your design look like it *could* work.

SUGGESTIVE REFERENCE

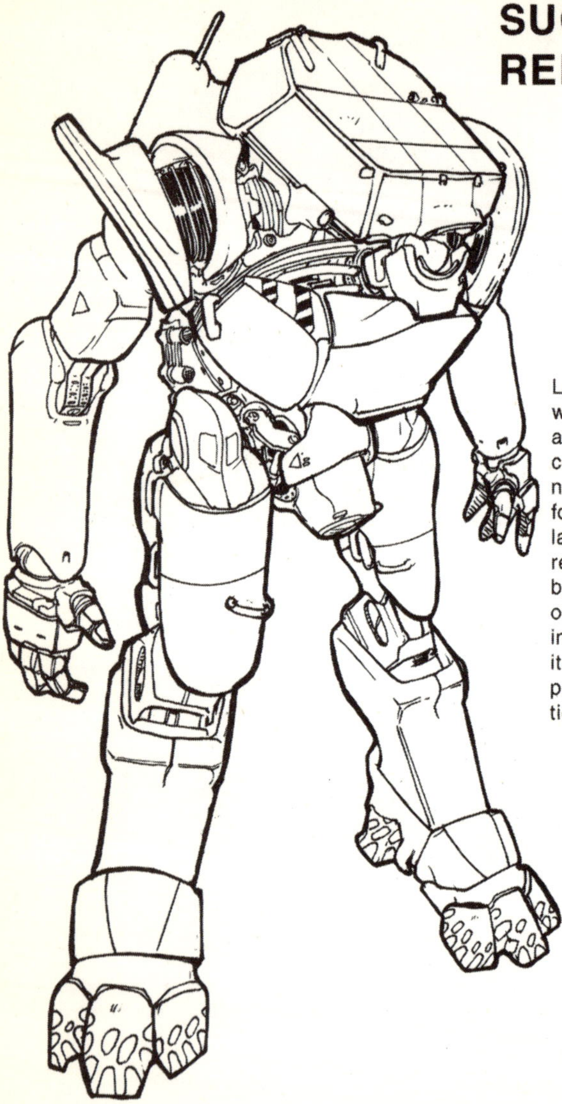

Let's apply some of what we've just talked about with this urban combat mech. There's no background and no foreground, so how large is it? You can't really tell, right? Well, by adding certain objects but not changing the size of the mech itself, you'll see the power of visual suggestion.

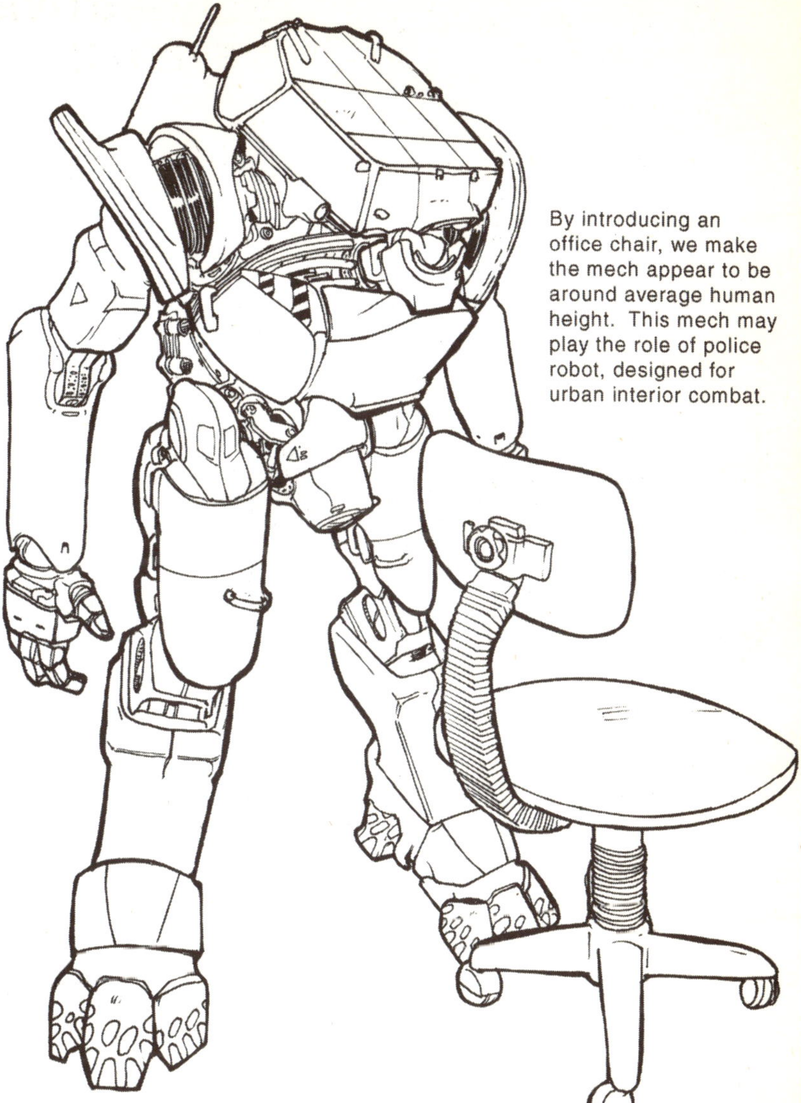

By introducing an office chair, we make the mech appear to be around average human height. This mech may play the role of police robot, designed for urban interior combat.

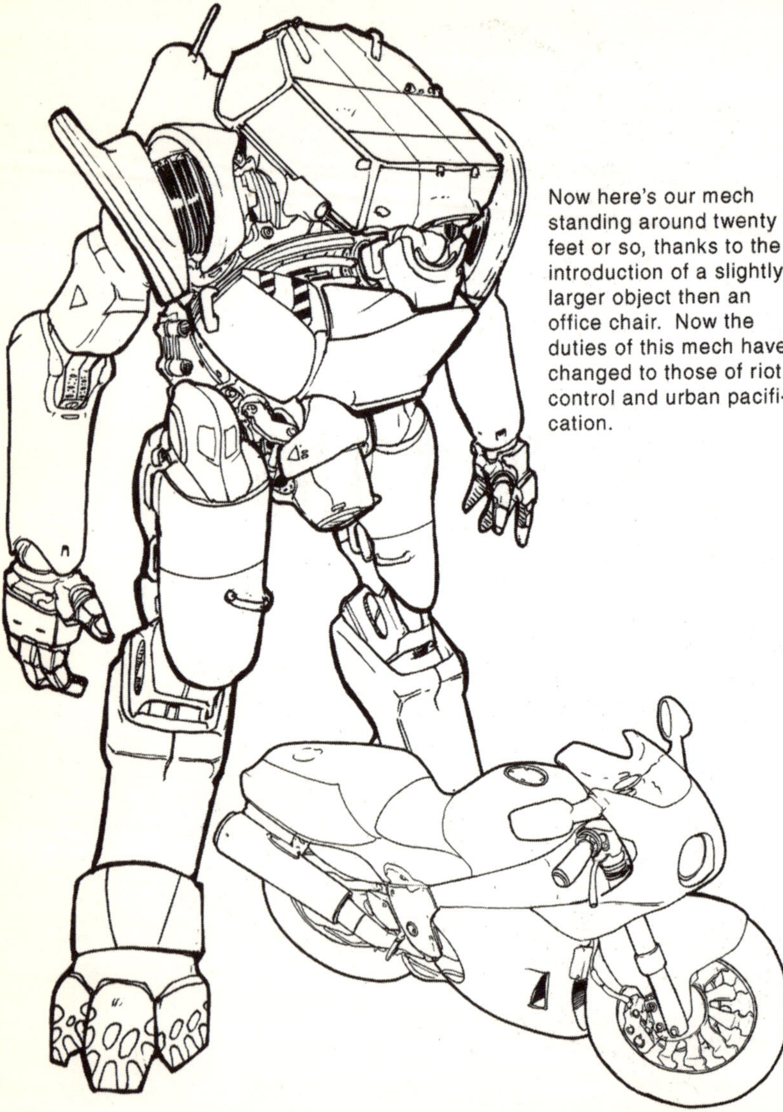

Now here's our mech standing around twenty feet or so, thanks to the introduction of a slightly larger object then an office chair. Now the duties of this mech have changed to those of riot control and urban pacification.

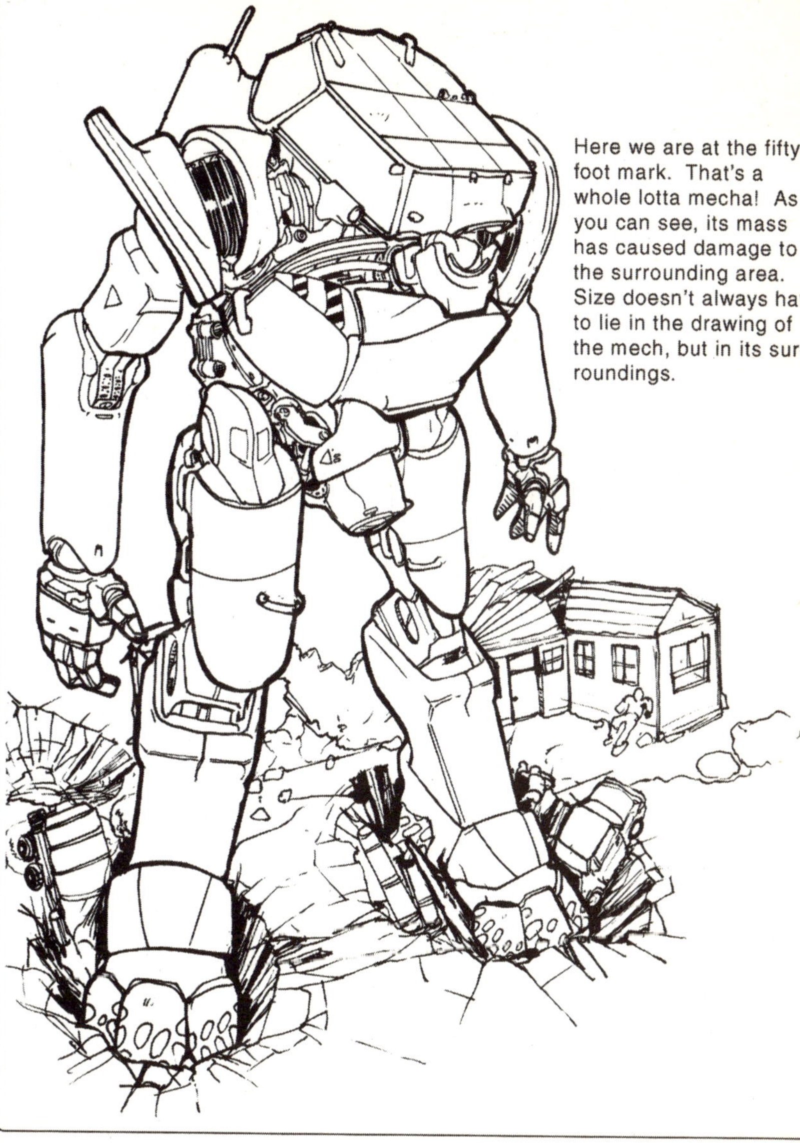

Here we are at the fifty-foot mark. That's a whole lotta mecha! As you can see, its mass has caused damage to the surrounding area. Size doesn't always have to lie in the drawing of the mech, but in its surroundings.

The size of mecha not only affects the mecha themselves but the environment as well. When a human-sized robot bashes a car, it crushes the top and collapses the suspension. Lots of damage, but not a total write-off.

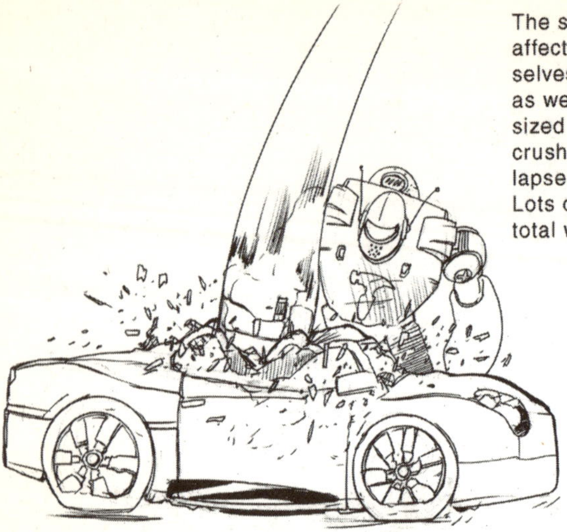

Larger robot, same car. The results are a little different. Because the larger robot has more mass and a larger surface area on the fists, more of the car is covered and ultimately flattened.

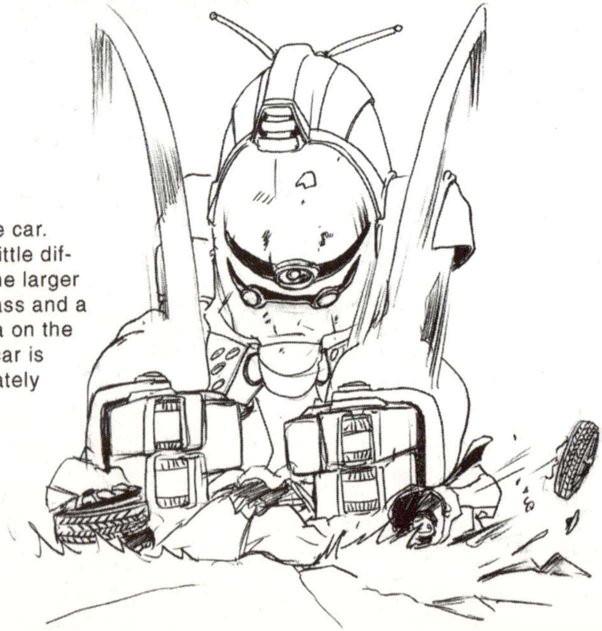

Then there's the giant robot. The weight of the mecha affects not just the car and the immediate area, but the entire surrounding environment.

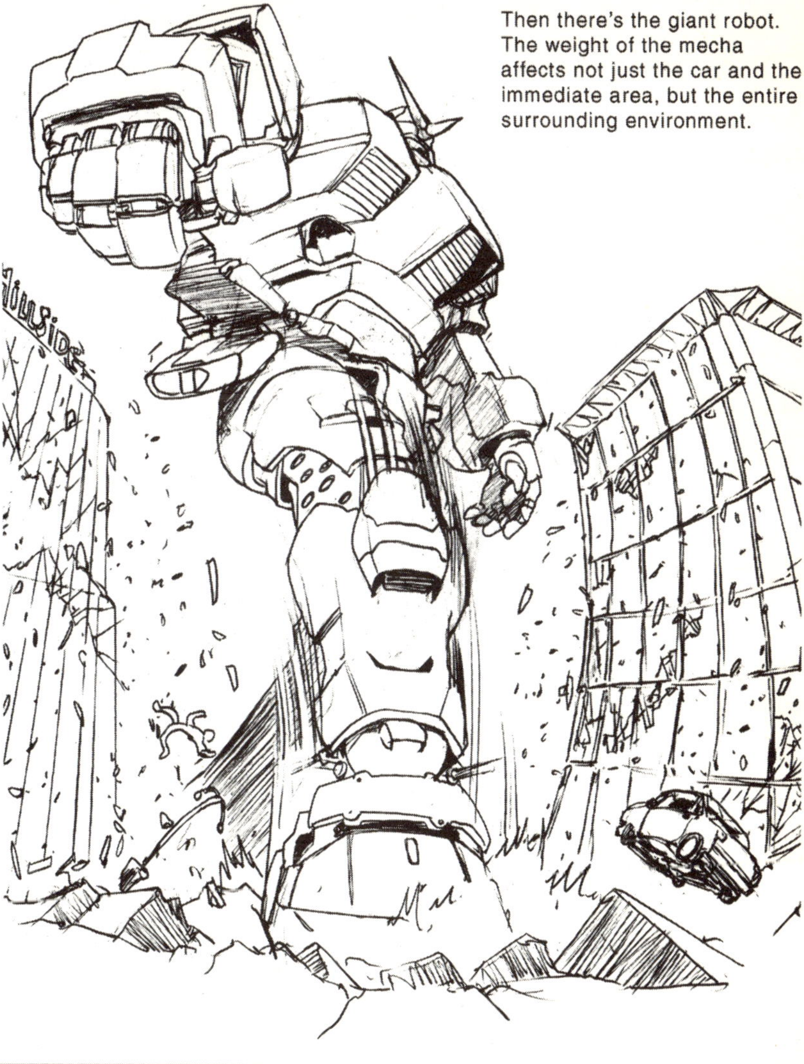

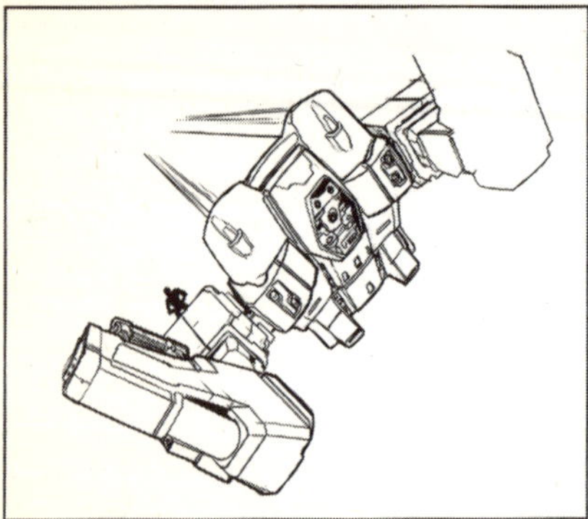

On the previous pages, we learned how a mech's size can be determined by its environment, and that this helps imply mass.

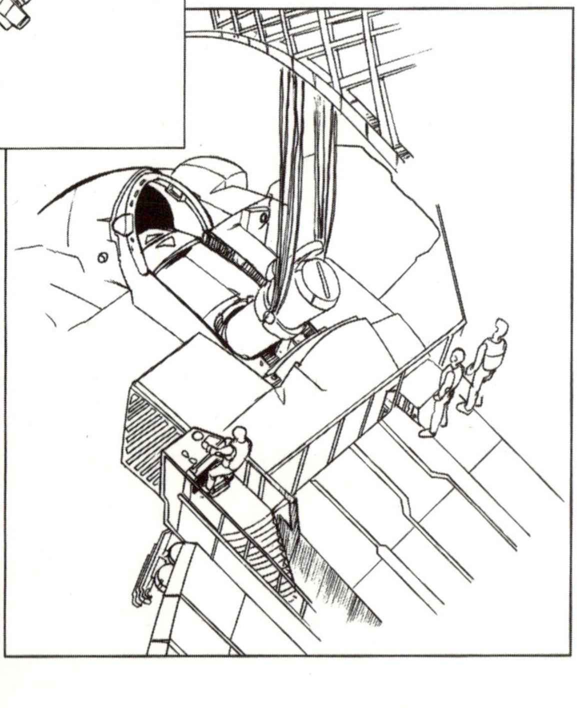

Using what you've learned about size relations, you can proceed to plot out an action sequence.

BUILDING UP A COMBAT SCENE

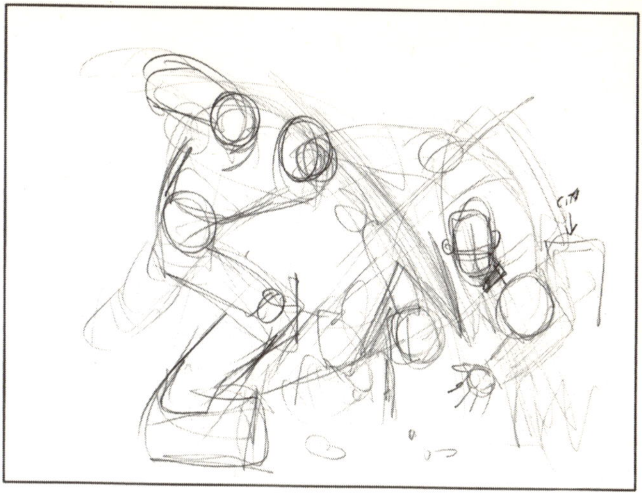

Start off with the rough outline of the forms.

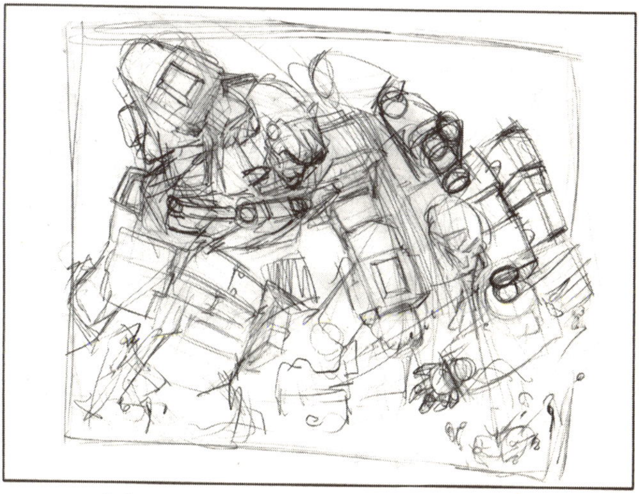

Build up from outline. Define point of impact (punch).

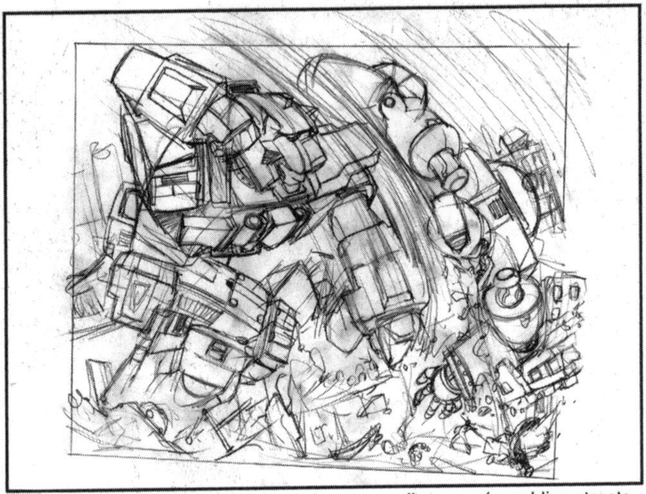

Show impact and residual effect on the surrounding area (crumbling streets, shattered windows...).

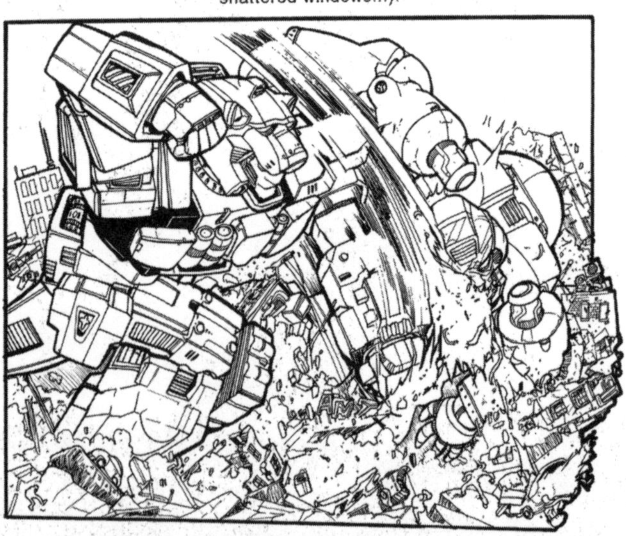

Add inks to clarify the action and clean up the piece.

CHAPTER 04

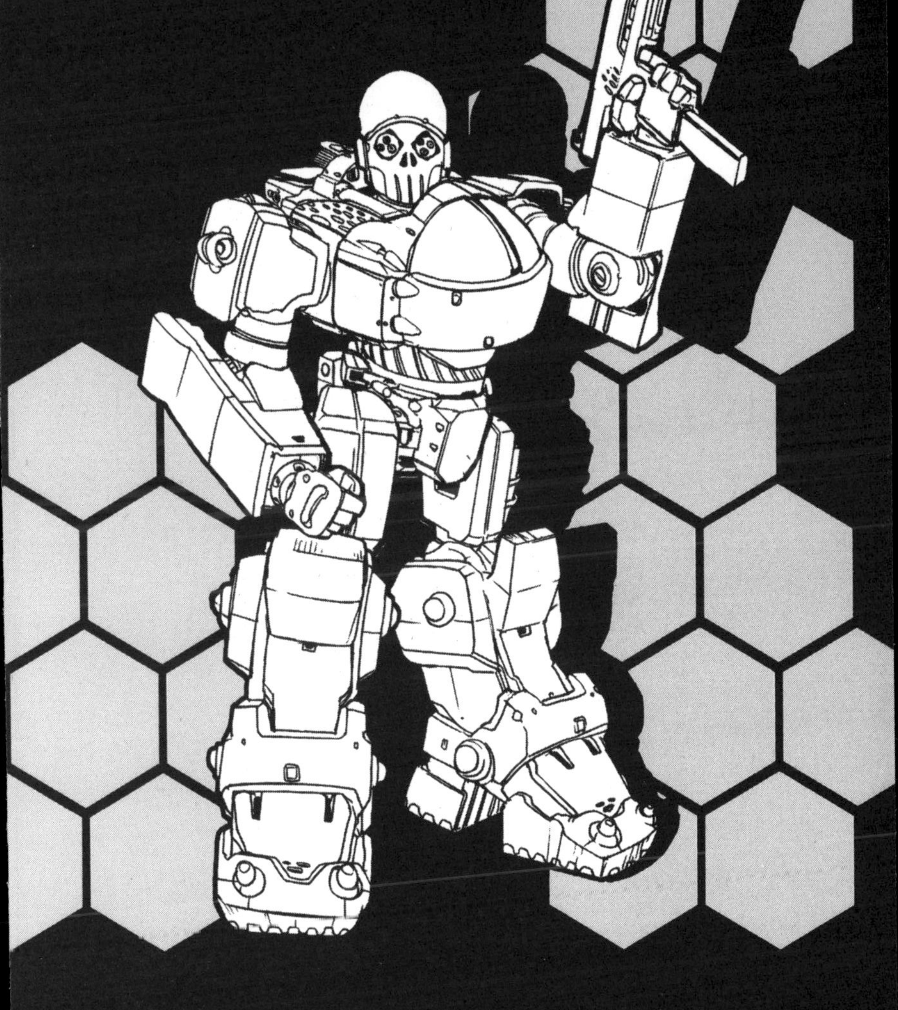

COMBAT MECHA

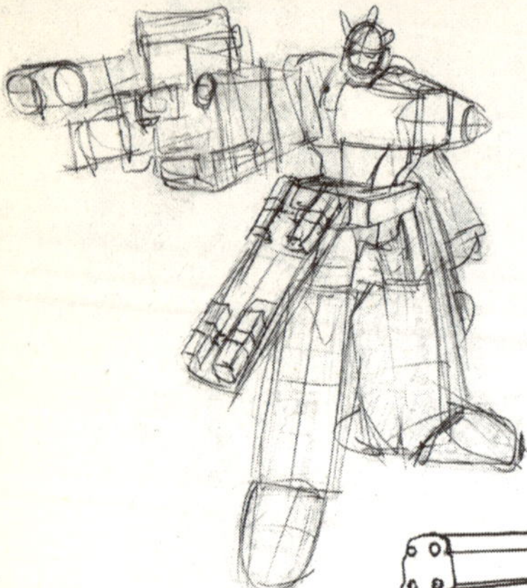

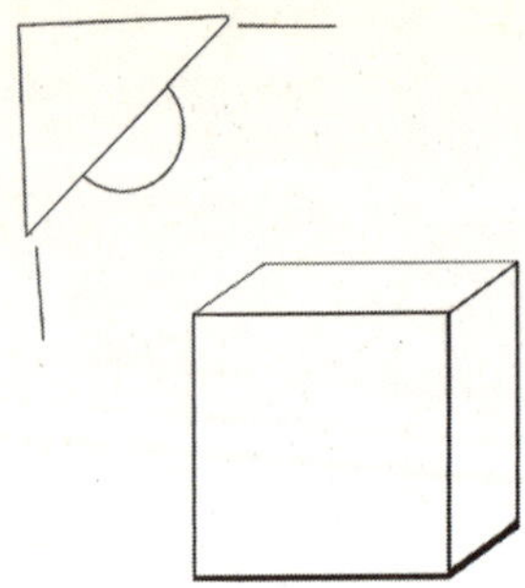

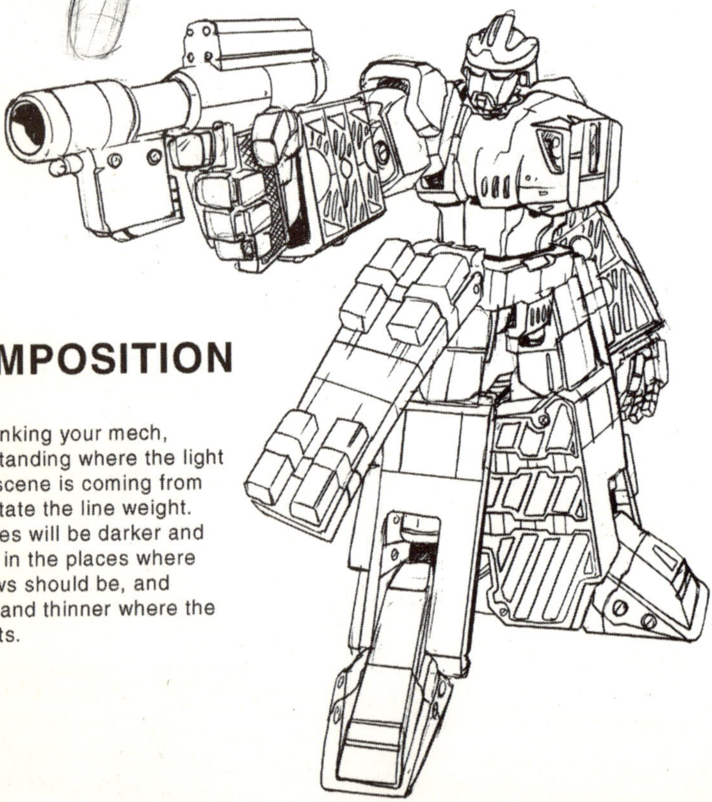

COMPOSITION

When inking your mech, understanding where the light in the scene is coming from will dictate the line weight. The lines will be darker and thicker in the places where shadows should be, and lighter and thinner where the light hits.

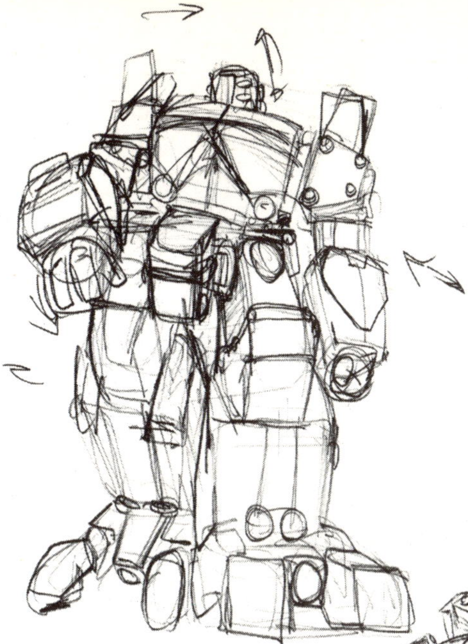

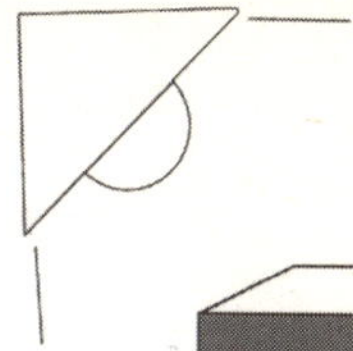

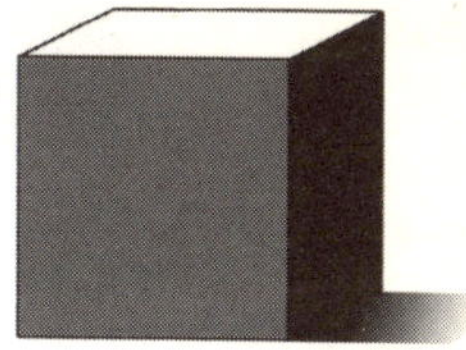

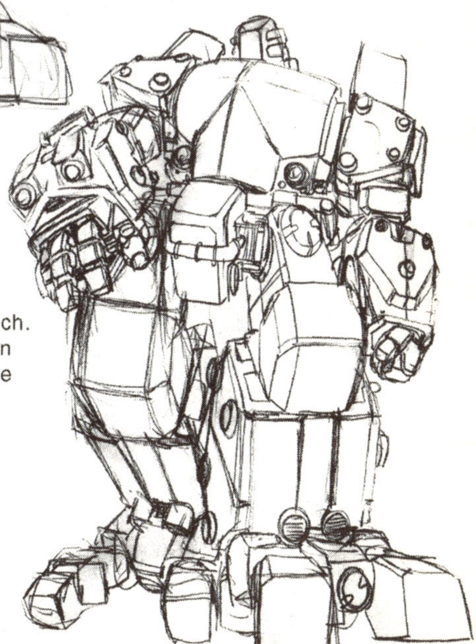

Keep track of the angles on the mech. The way an arm or leg is turned can make a drastic difference in how the shadows fall across its body.

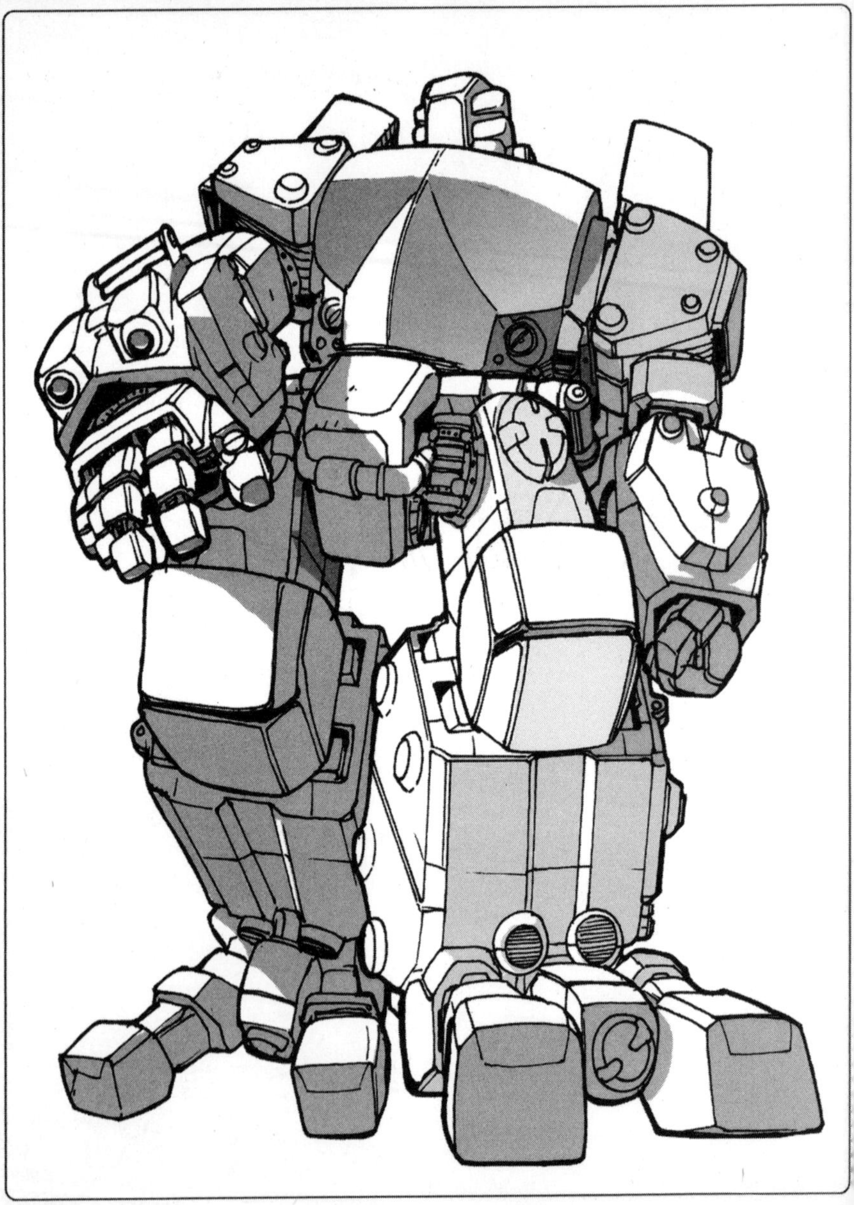

UNDERSTANDING THE FORM

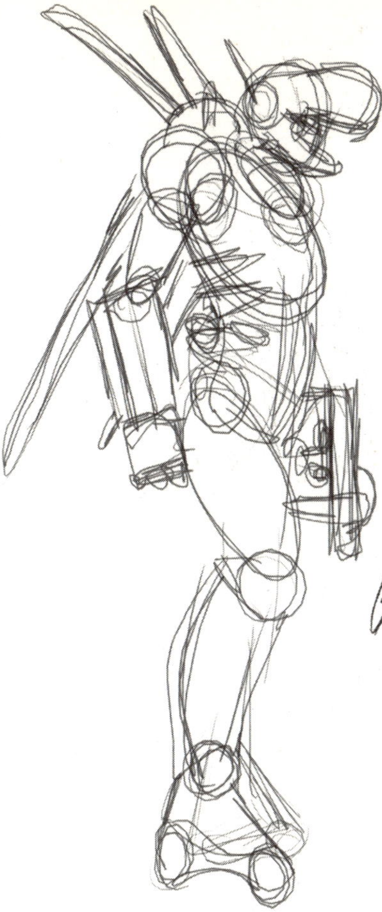

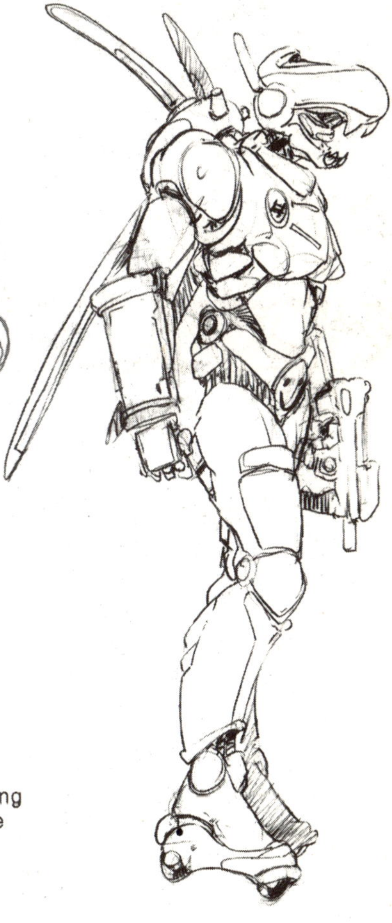

Designing more organic-looking mecha is an exercise in figure drawing rather than technical drawing.

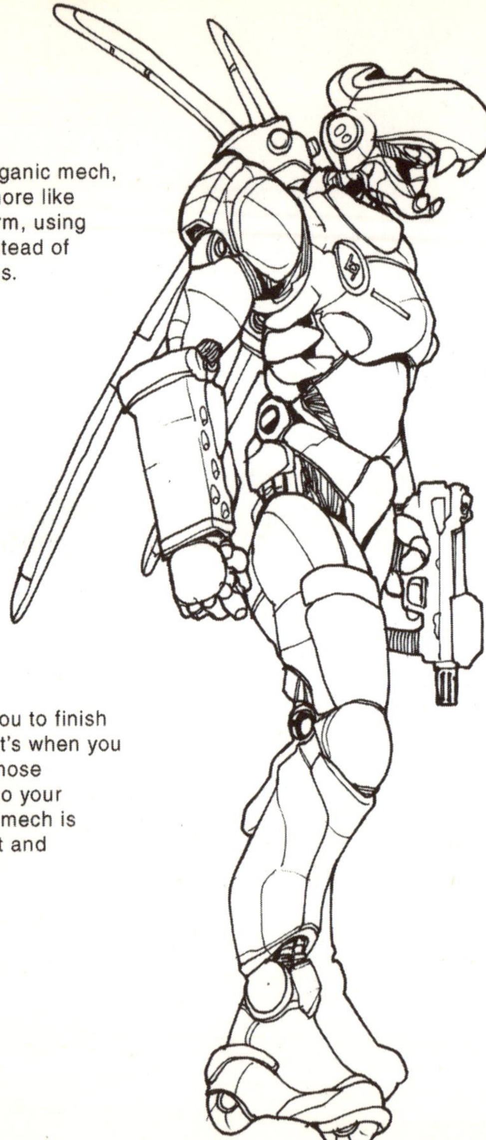

When drawing an organic mech, your layout will be more like drawing a human form, using circles and ovals instead of cubes and rectangles.

When it's time for you to finish off the drawing, that's when you can start to apply those "robotic" elements to your organic form. This mech is designed to be light and extremely mobile.

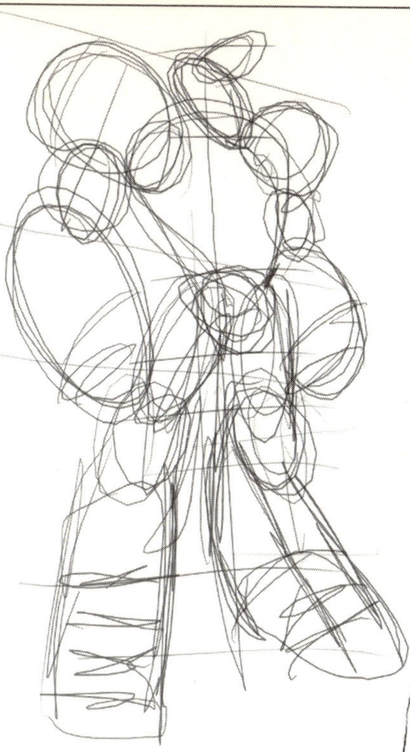

The last mech had the look and build of a stealth or assassin unit. Organic combat mecha follow the same basic principles using a different base form.

The basic layout for an organic combat mecha is to take a stocky human form and begin to add weapons and gear. The mecha has a fluid form because the elements follow the contours of its body.

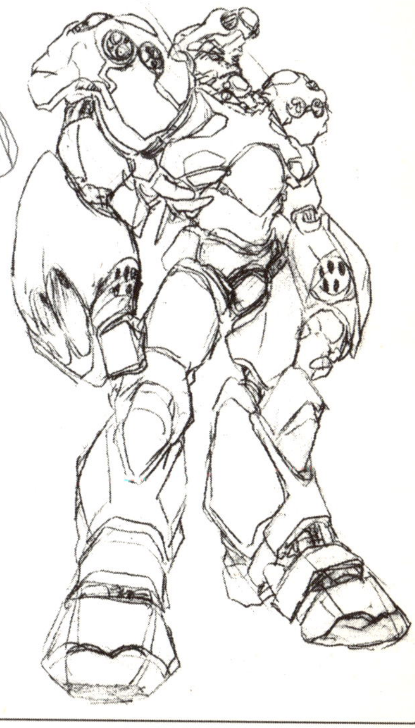

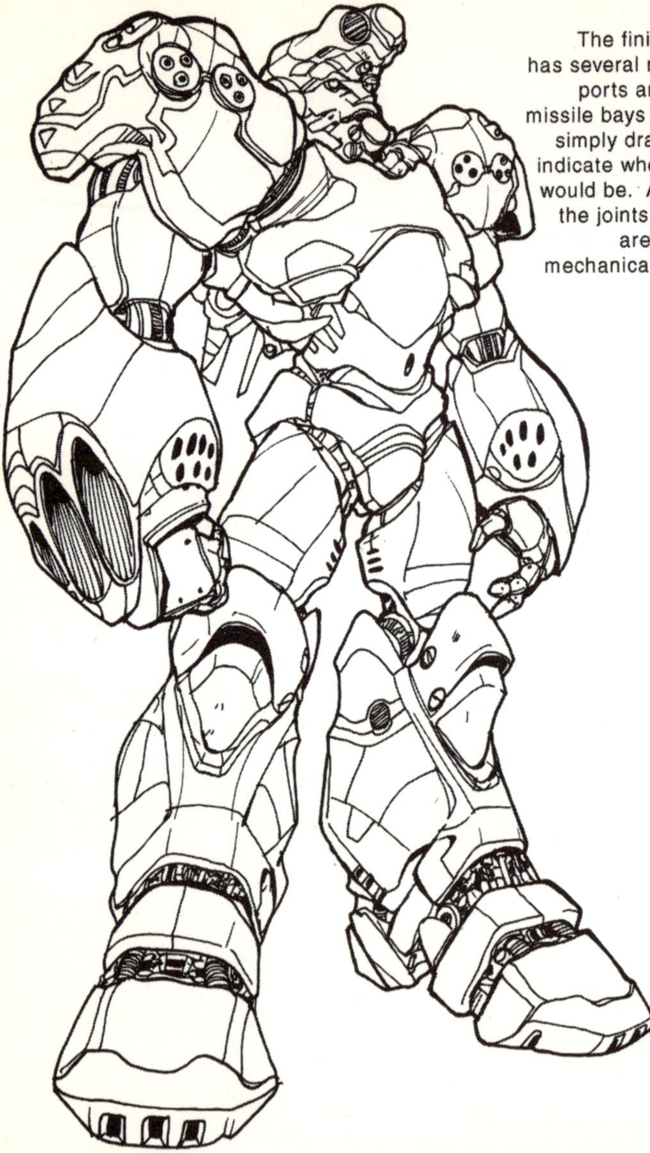

The finished product has several recessed gun ports and concealed missile bays illustrated by simply drawing lines to indicate where the doors would be. Also, many of the joints and muscles are mimicked by mechanical equivalents.

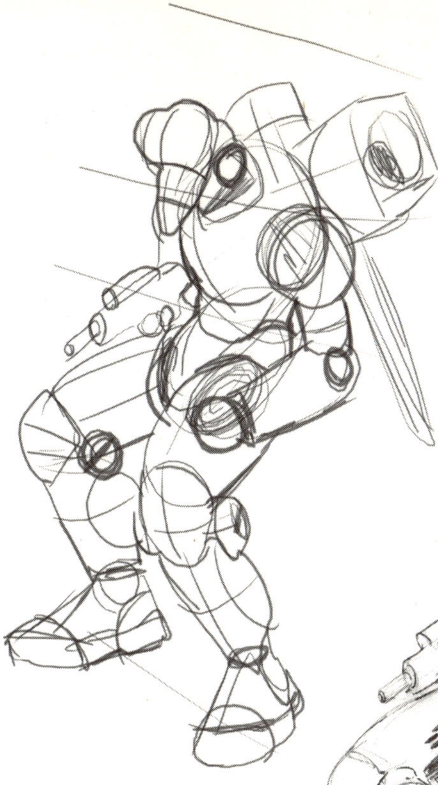

This mech is designed for heavy assault. The base form chosen here is that of a combat marine: light armor, a reliable light weapon, and a very open cockpit/camera area.

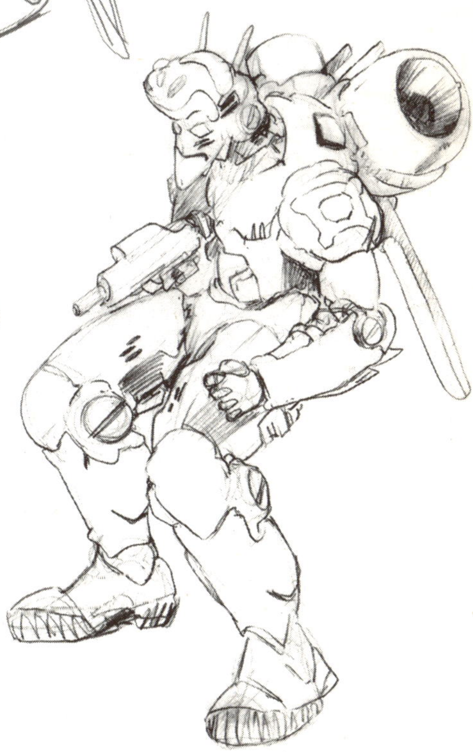

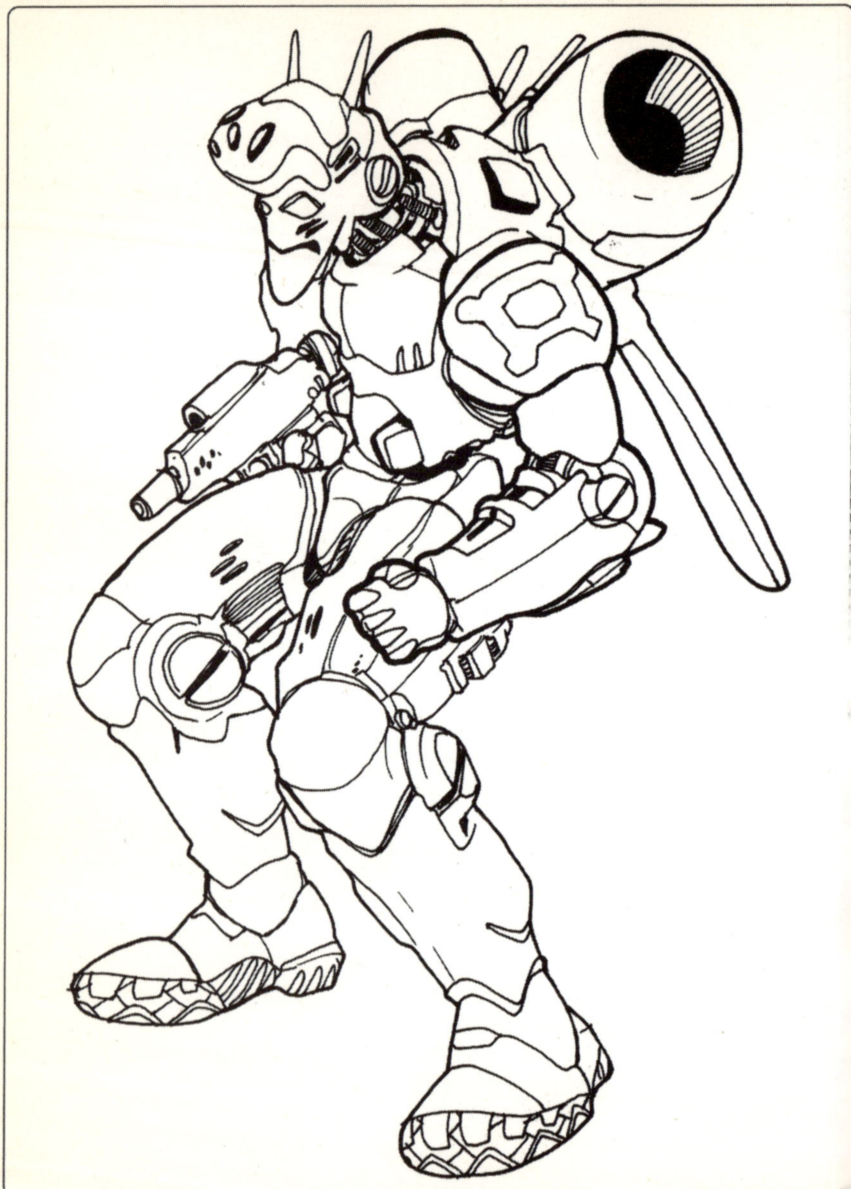

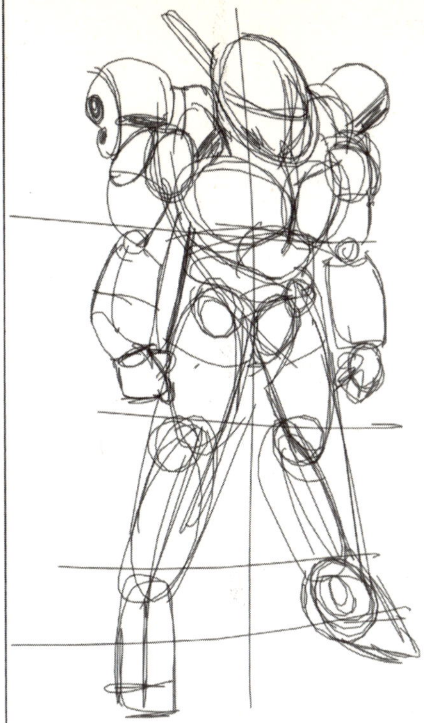

Make sure you maintain a sense of clarity in your design. No matter how stylish you make your piece, you want to make it easily understandable to the viewer.

It's difficult to make out certain elements in the illustration to the right. For example, the backpack and the arms seem to run into each other.

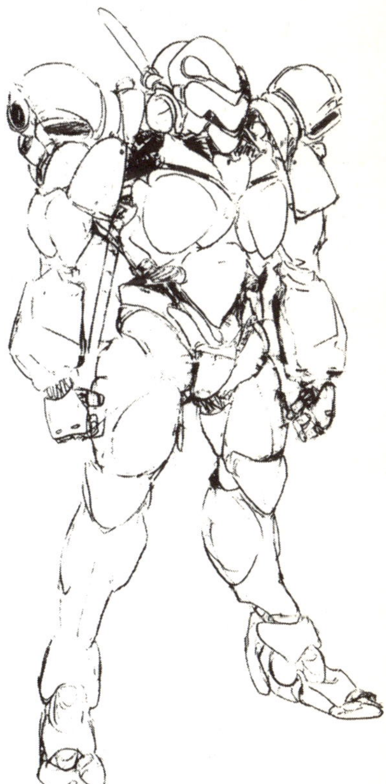

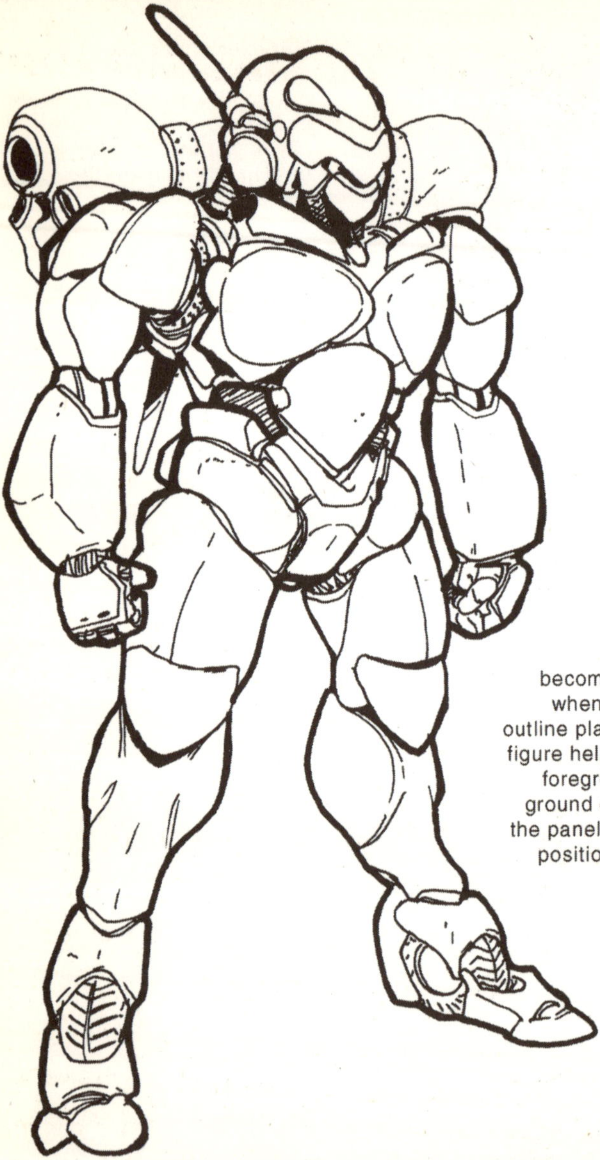

The illustration becomes a lot clearer when it's inked. The outline placed around the figure helps separate the foreground and background elements, while the paneling clarifies the position of the arms.

ADDITIONAL SUPPORT

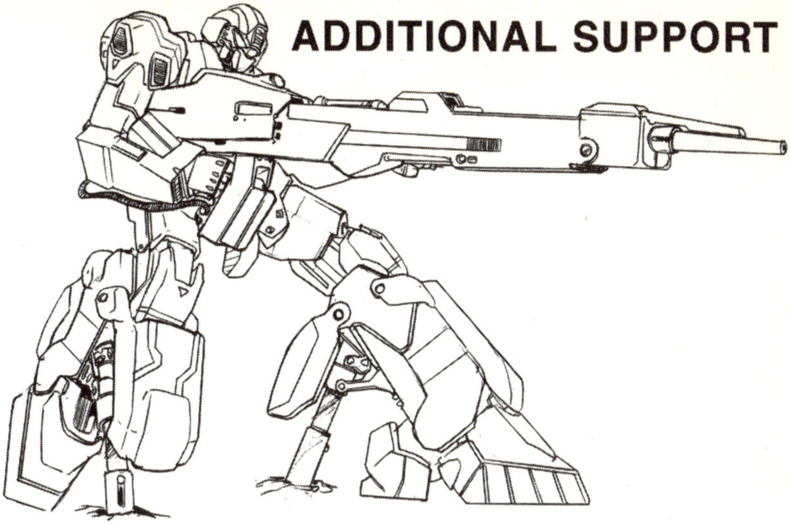

This mech has built-in braces to support its body against the recoil of the gun. You can use a mech's design to explain practical applications.
For example, if the mech has wide legs, you can say it has built-in boosters that are hidden by its wide outer shell. In this case, the wide legs hide the stabilizing rods.

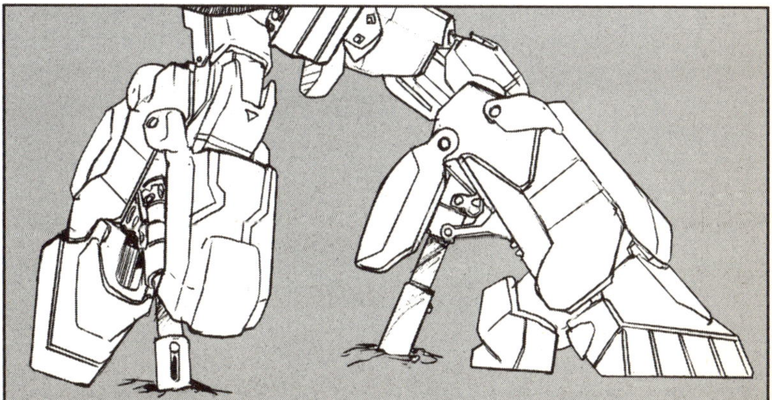

COCKPIT SETUP

Cockpits in mecha are usually placed in the head. It's naturally the highest point of the body, and it offers the pilot a clear line of sight.

This mech has a similar setup to a tank. Its focus is on the heavy shoulder-mounted gun.

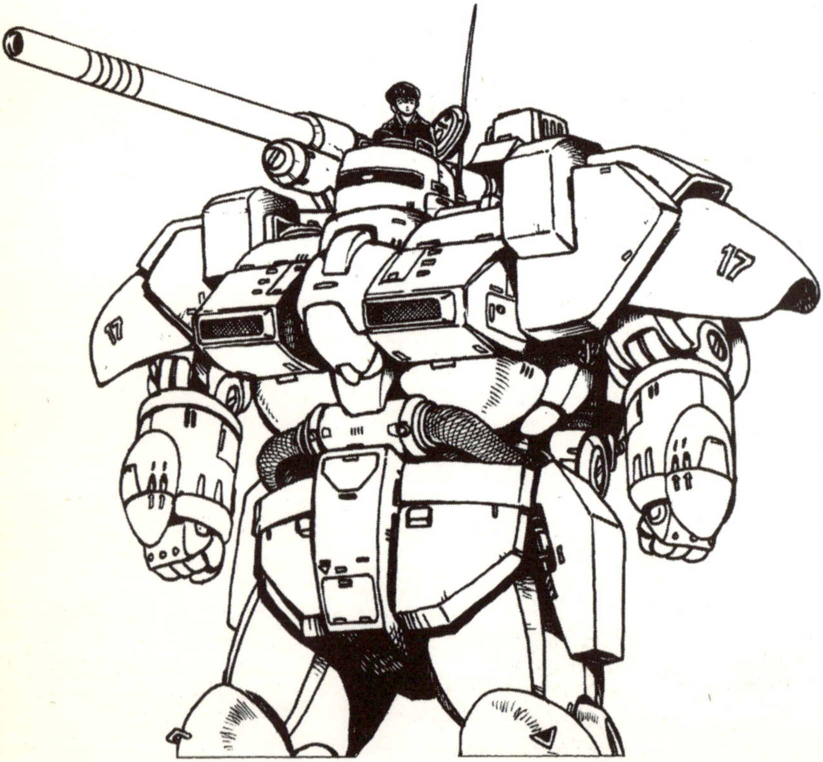

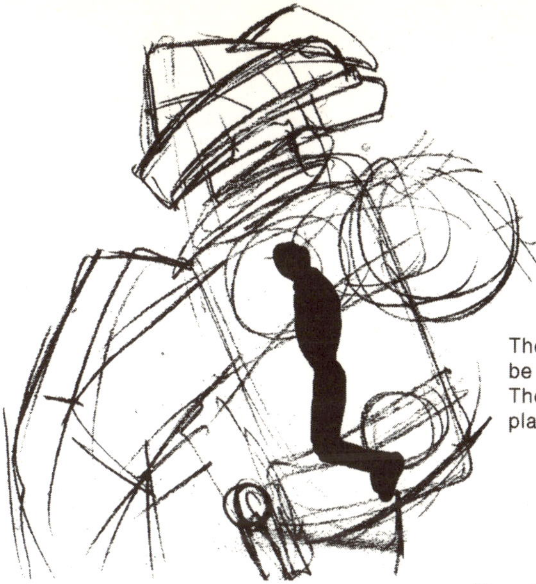

The cockpit could also be placed in the chest. The benefit of this placement is protection.

The pilot is positioned vertically inside a capsule-like cockpit. Instead of a simple stick-and-pedal setup, he could be in a harness that helps him control the movements of the mech.

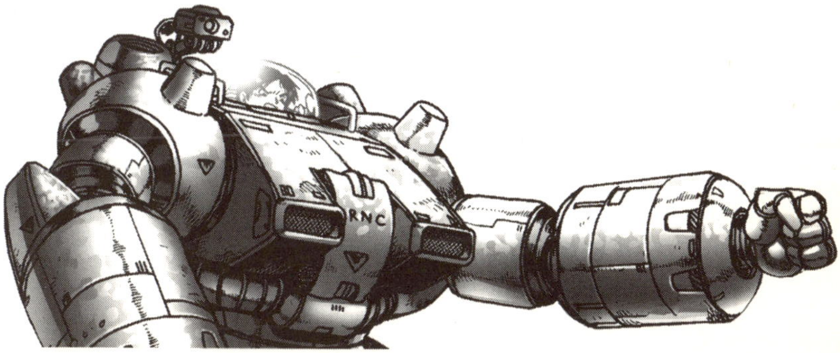

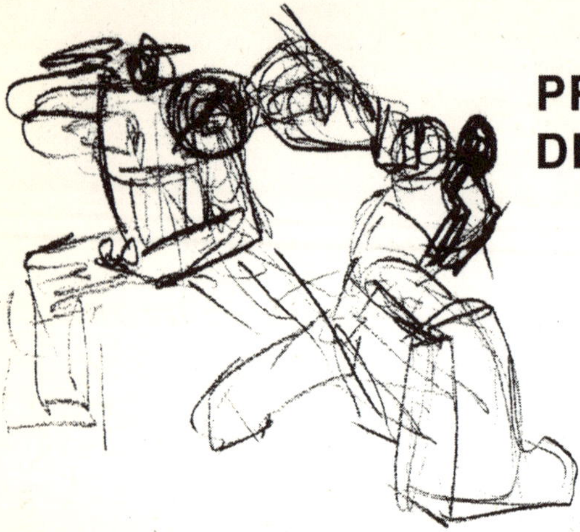

PRELIMINARY DESIGN STAGE

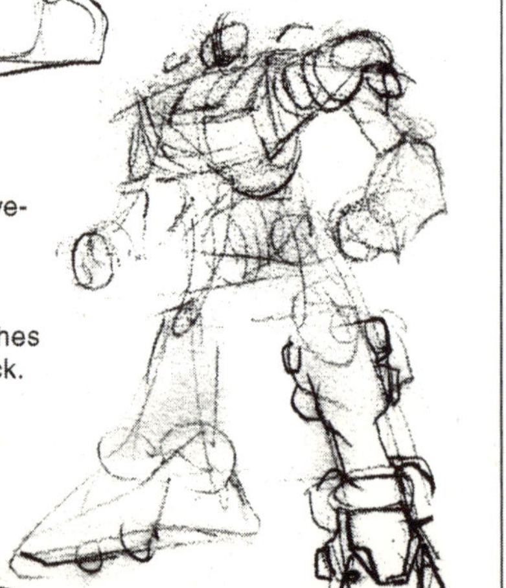

Drawing small allows you to plot out movement with an overly detailed mech.
It's an easy way to get some of the best poses without wasting too much time on rendering. Remember to keep the sketches small, simple, and most importantly, quick.

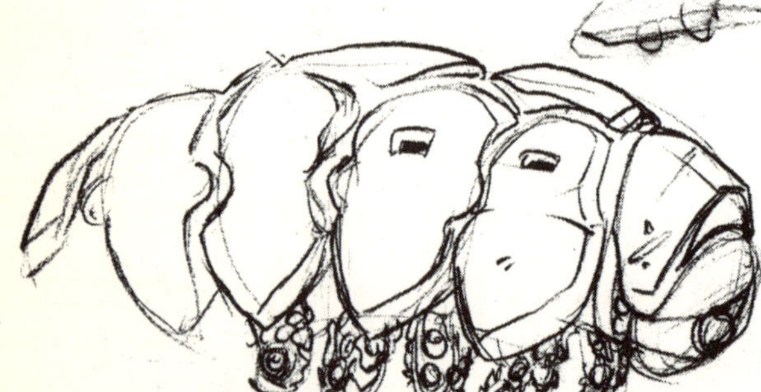

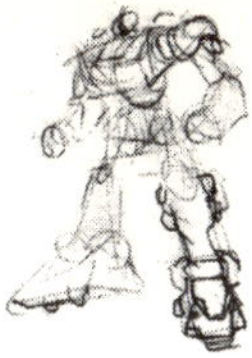

Once you've gotten a decent pose scribbled out, then you can start to render. The best thing about working this way is that you can start designing in one direction and end up with something completely different.

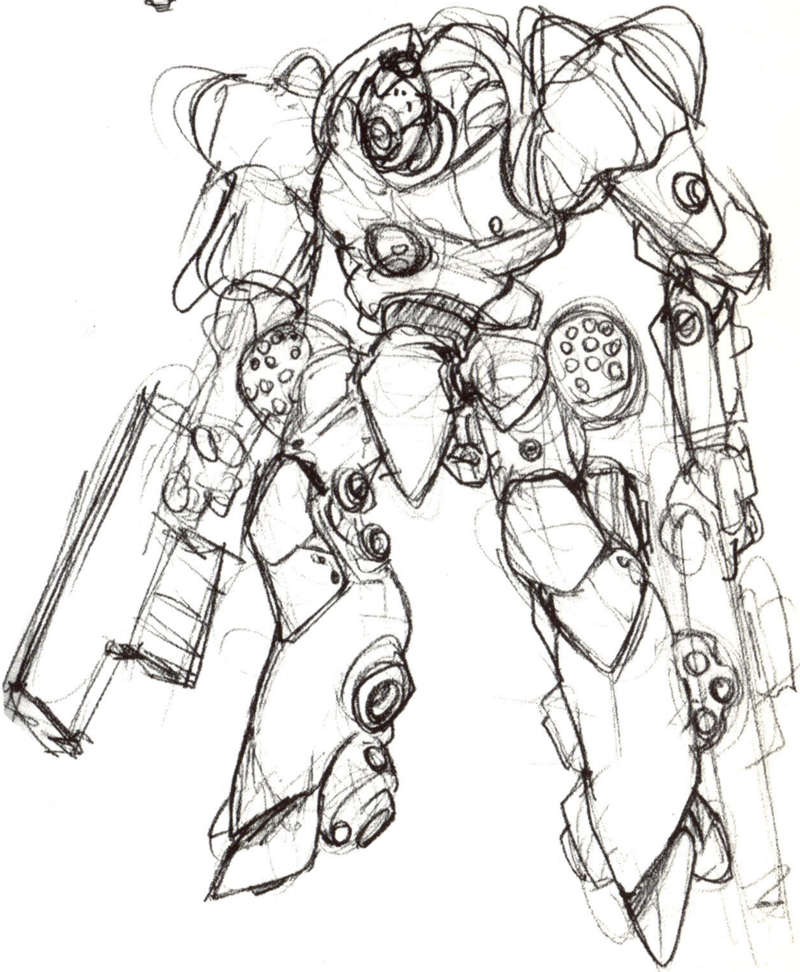

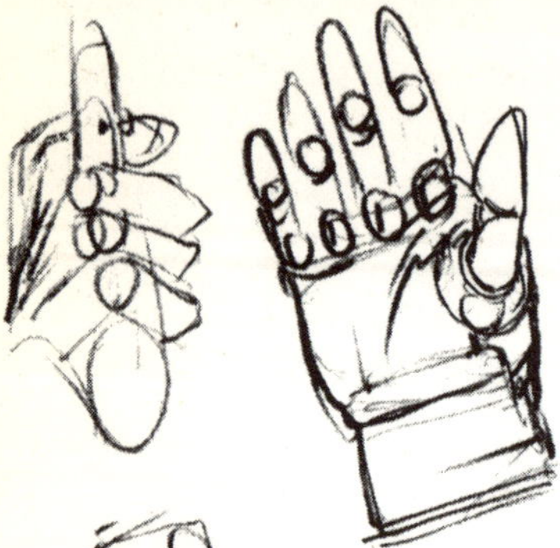

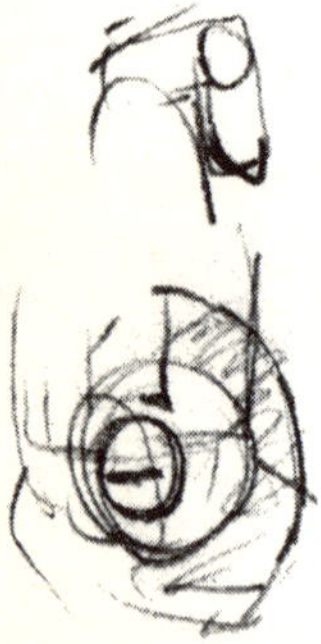

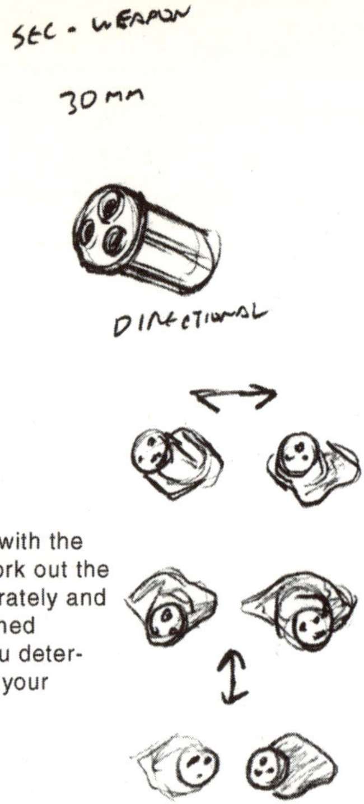

Another thing you can do with the preliminary sketches is work out the details of your mech separately and then add them to the finished drawing. This will help you determine the primary focus of your mech's design.

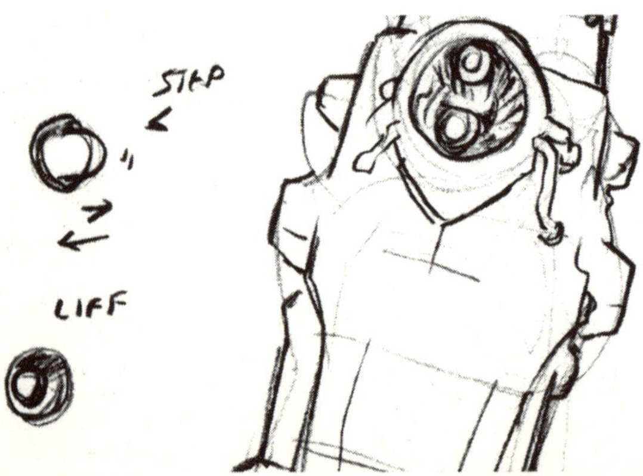

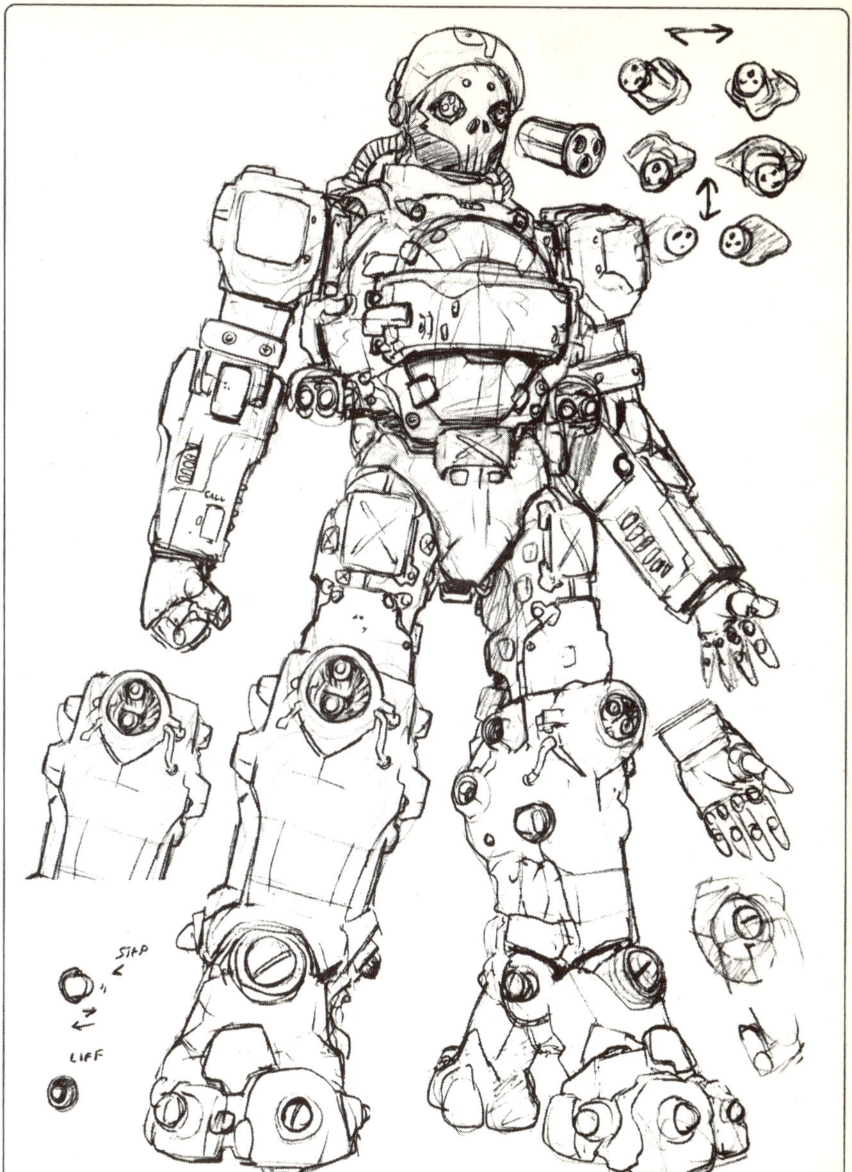
CALL

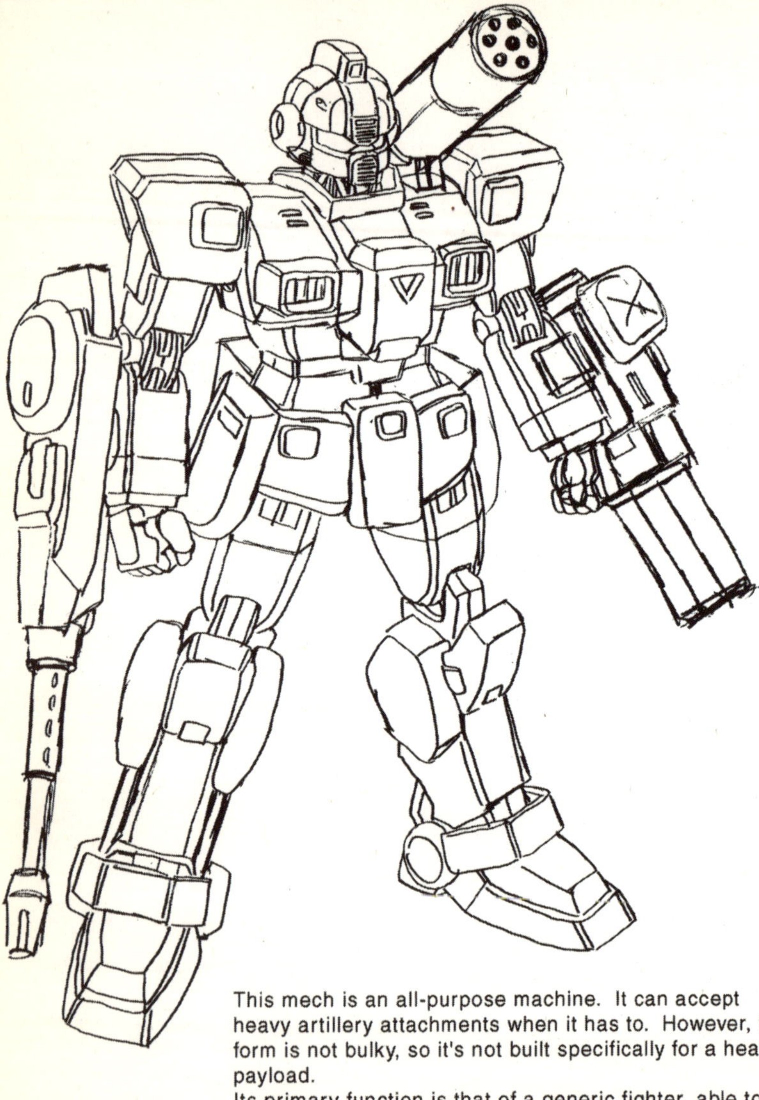

This mech is an all-purpose machine. It can accept heavy artillery attachments when it has to. However, the form is not bulky, so it's not built specifically for a heavy payload.
Its primary function is that of a generic fighter, able to pick up and drop its primary weapon.

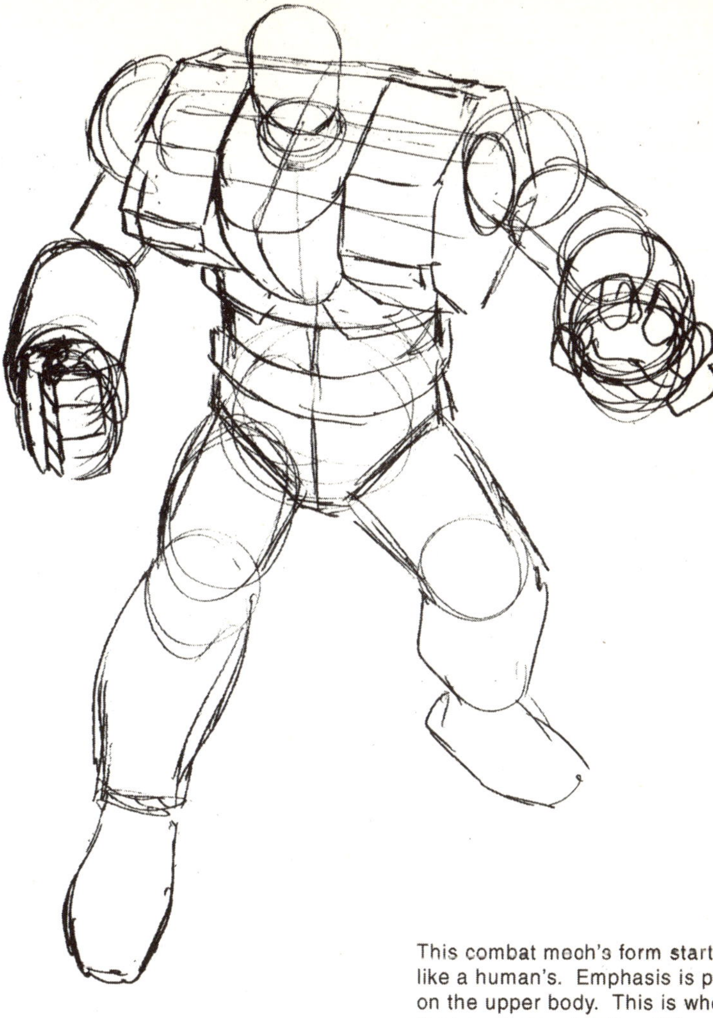

This combat mech's form starts off like a human's. Emphasis is placed on the upper body. This is where the key design details will be.

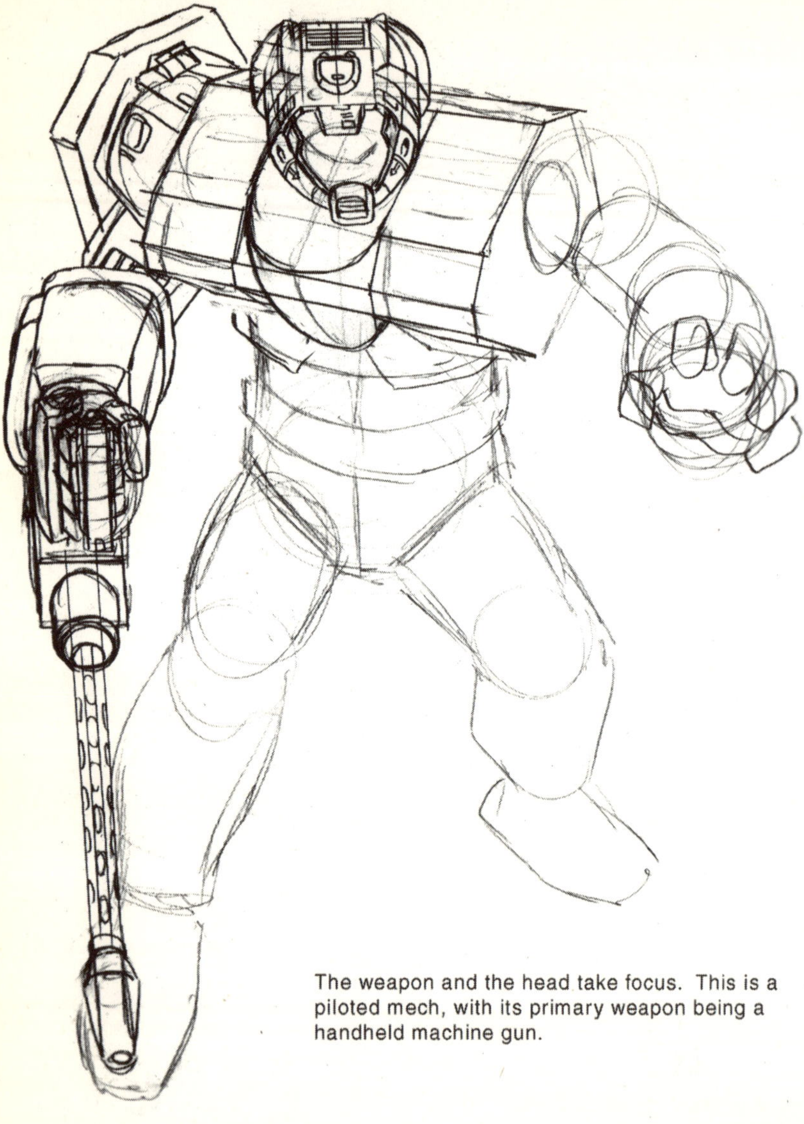

The weapon and the head take focus. This is a piloted mech, with its primary weapon being a handheld machine gun.

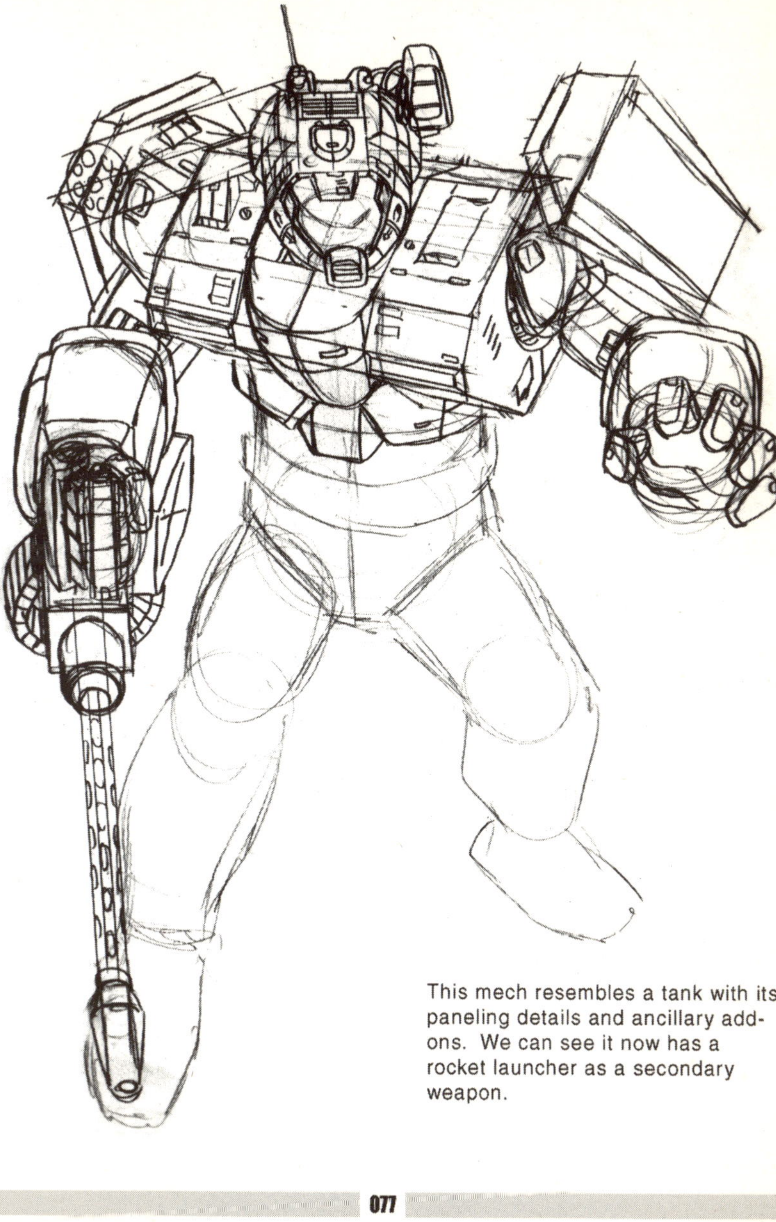

This mech resembles a tank with its paneling details and ancillary add-ons. We can see it now has a rocket launcher as a secondary weapon.

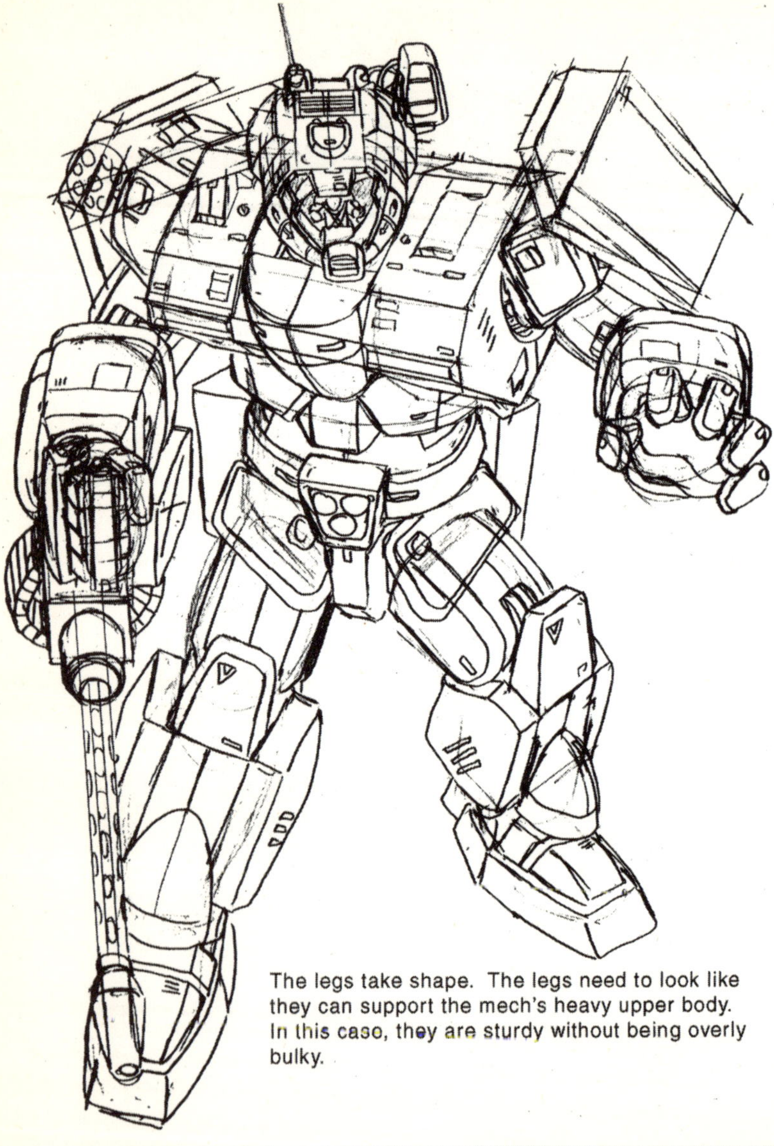

The legs take shape. The legs need to look like they can support the mech's heavy upper body. In this case, they are sturdy without being overly bulky.

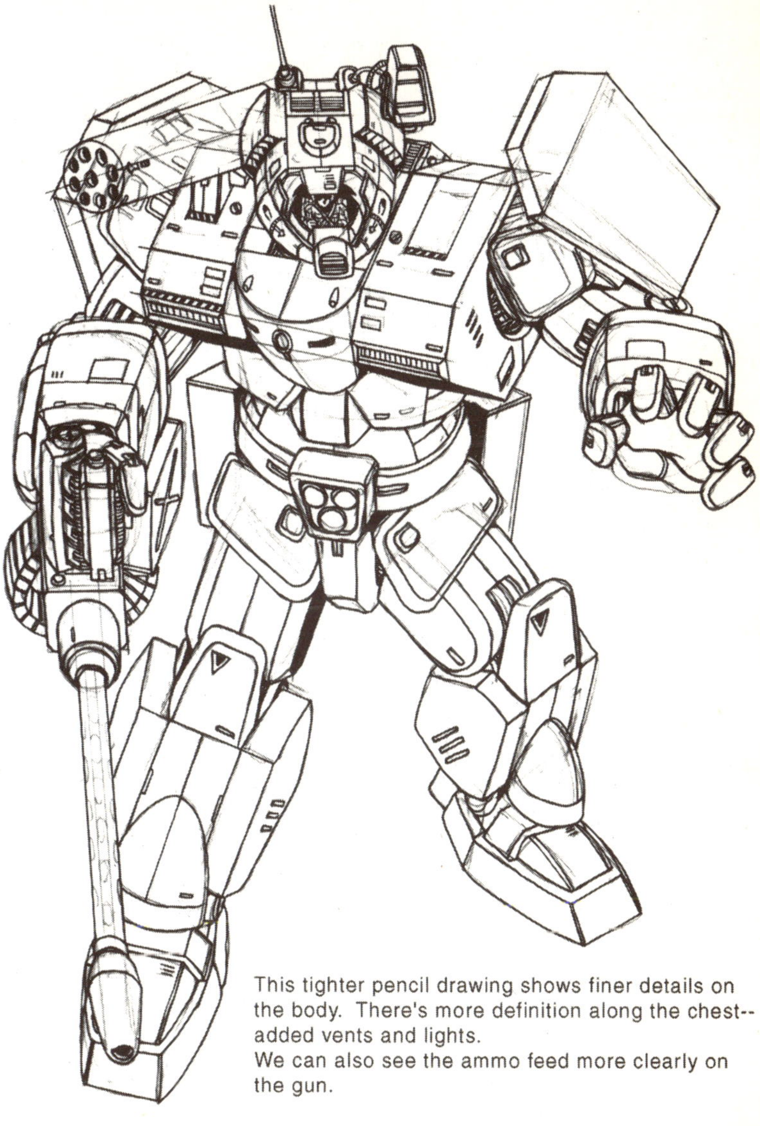

This tighter pencil drawing shows finer details on the body. There's more definition along the chest--added vents and lights.
We can also see the ammo feed more clearly on the gun.

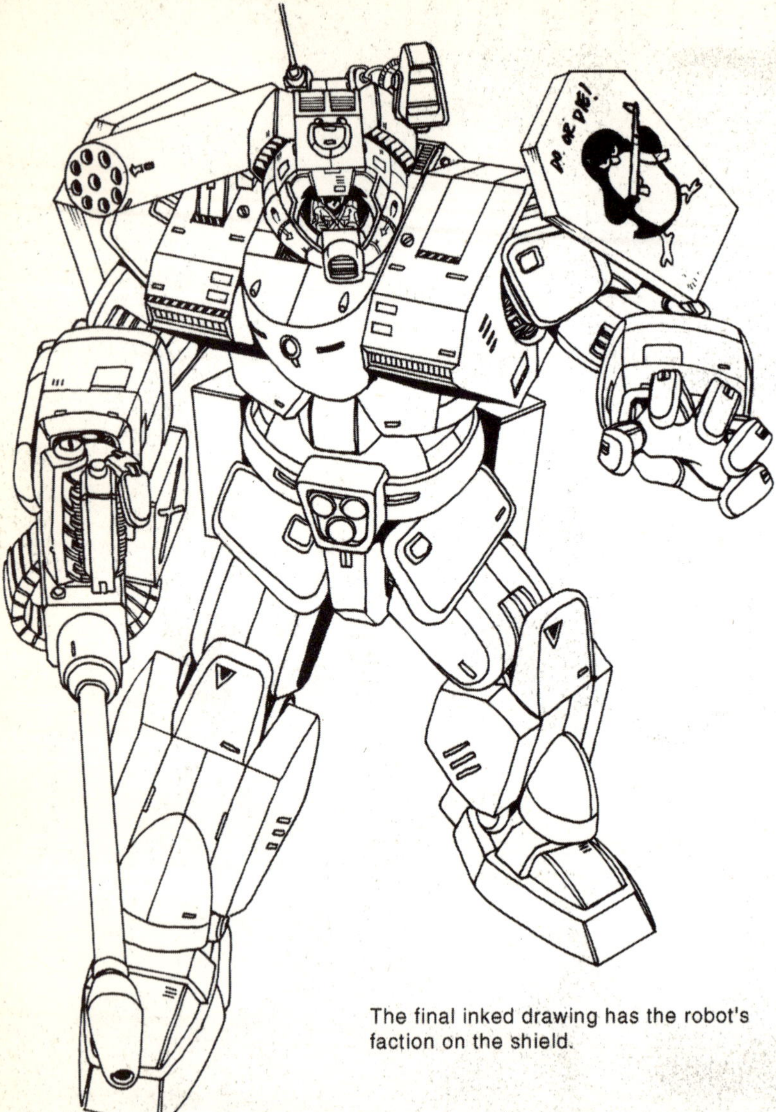

The final inked drawing has the robot's faction on the shield.

CHAPTER 05

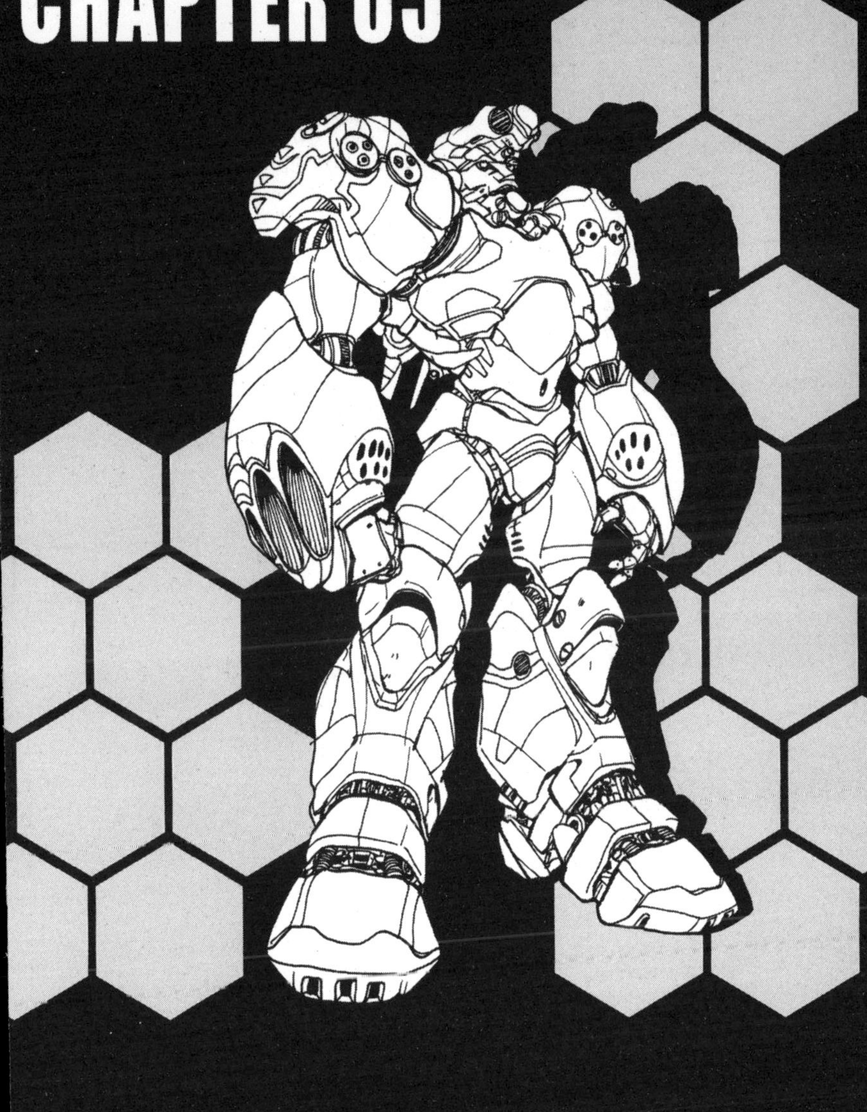

HEROIC MECHA

HEROIC MECHA

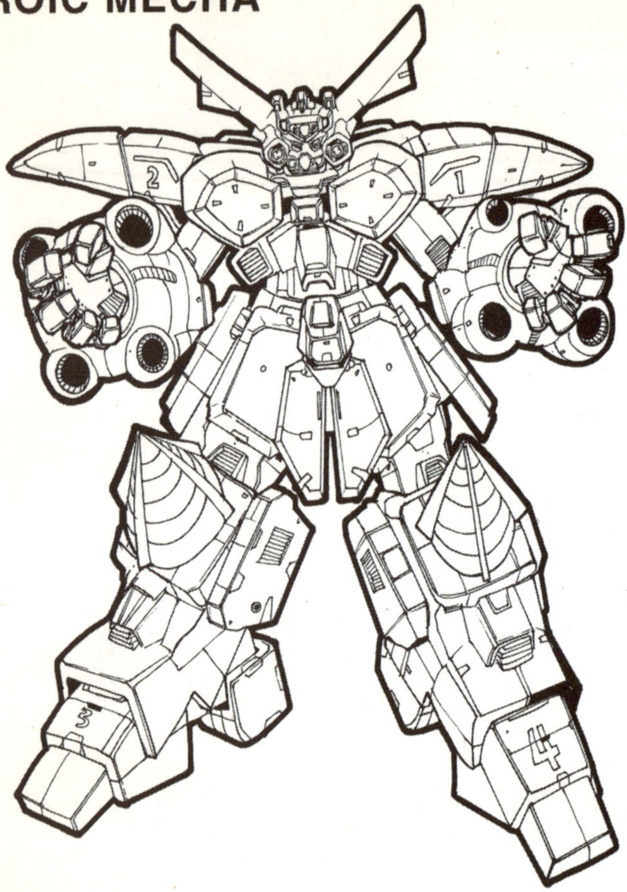

Heroic mecha are the equivalent of superheroes. They have special powers and their adventures are usually them fending off otherworldly monsters.

POWER-UPS

Part of the heroic mech's arsenal is power-ups. Here's a step-by-step.

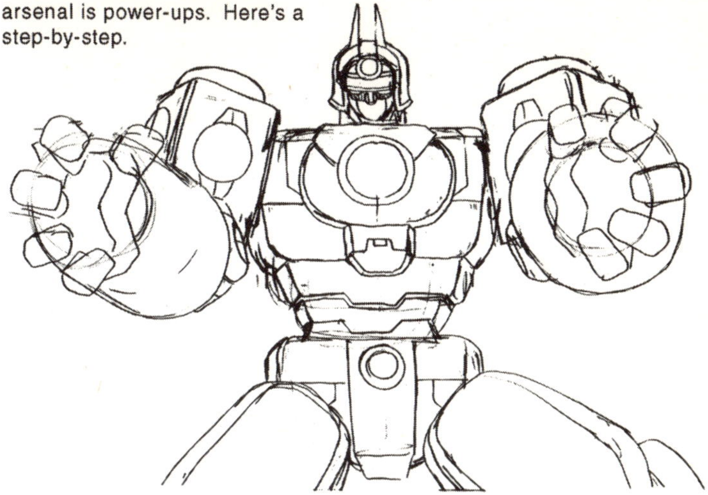

Step 1: Draw the pose for the power-up.

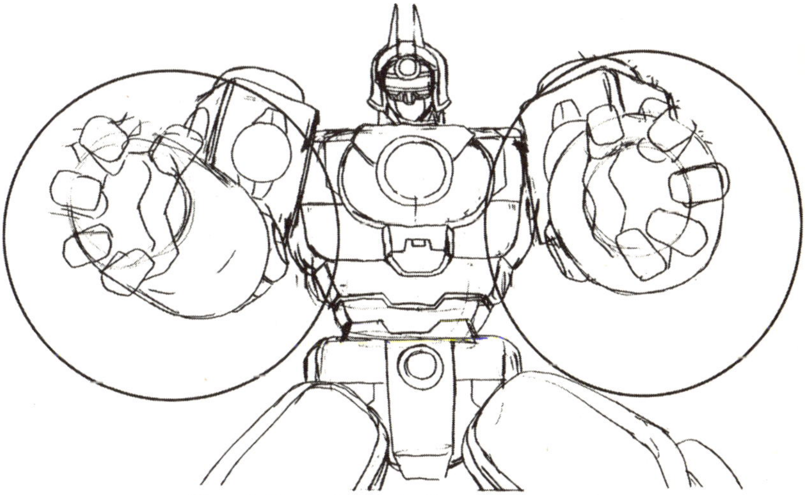

Step 2: Draw circles for the power-up radius.

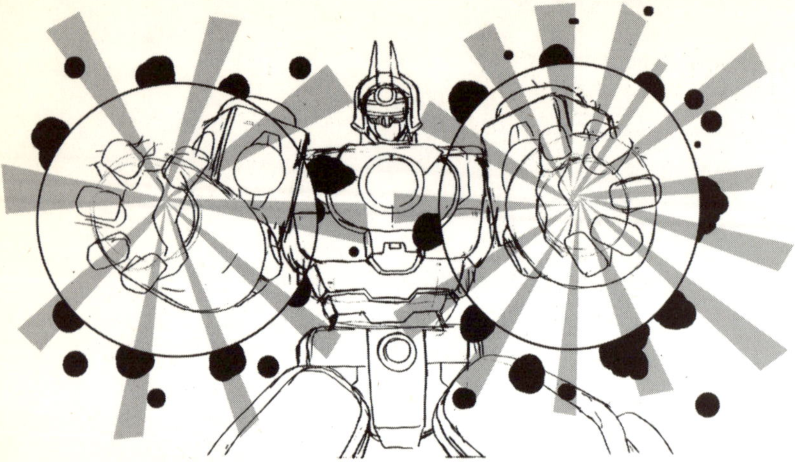

Step 3: Add the effects. Make sure the blast radiates from the center of the effect.

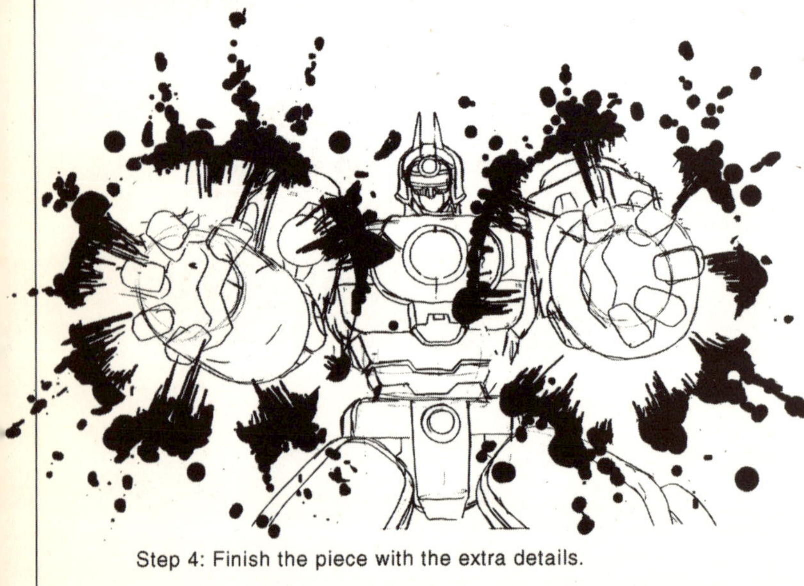

Step 4: Finish the piece with the extra details.

PILOTED FIGHTER

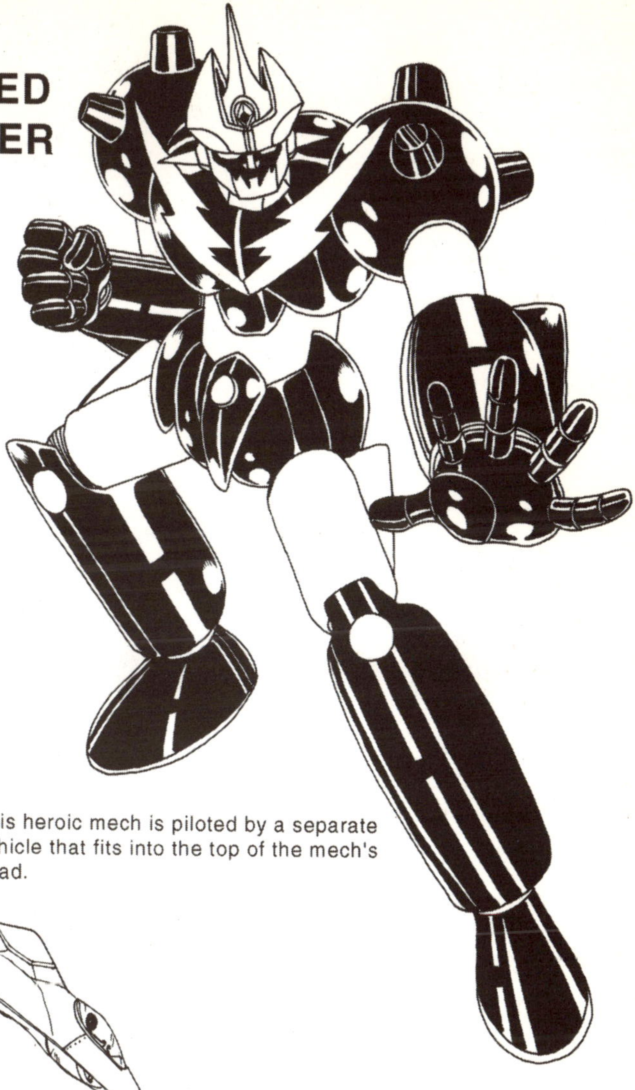

This heroic mech is piloted by a separate vehicle that fits into the top of the mech's head.

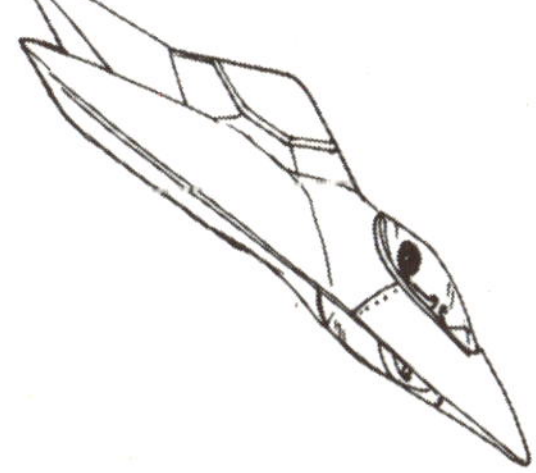

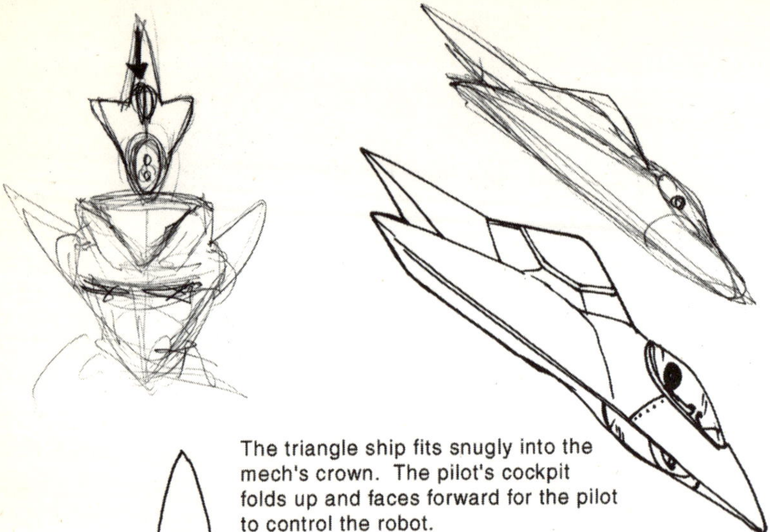

The triangle ship fits snugly into the mech's crown. The pilot's cockpit folds up and faces forward for the pilot to control the robot.

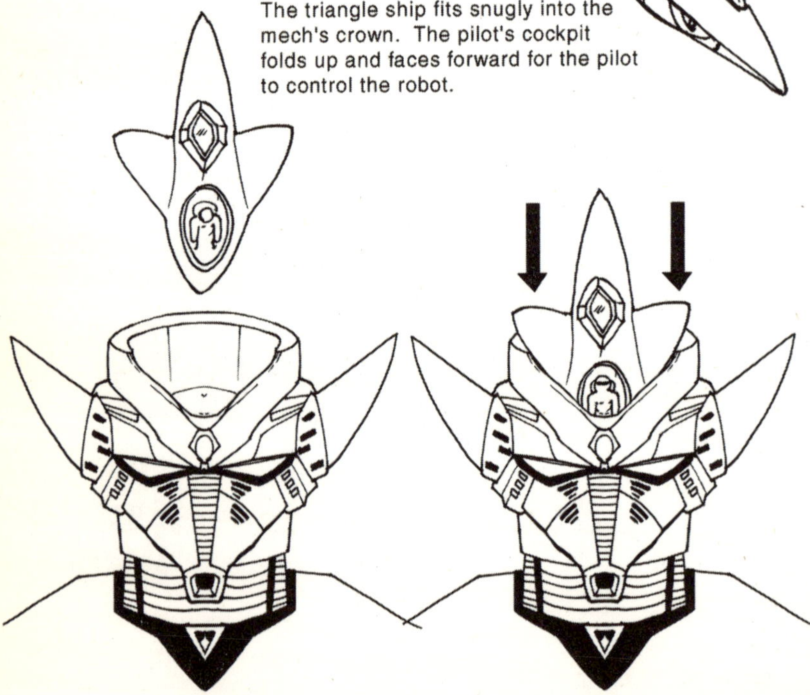

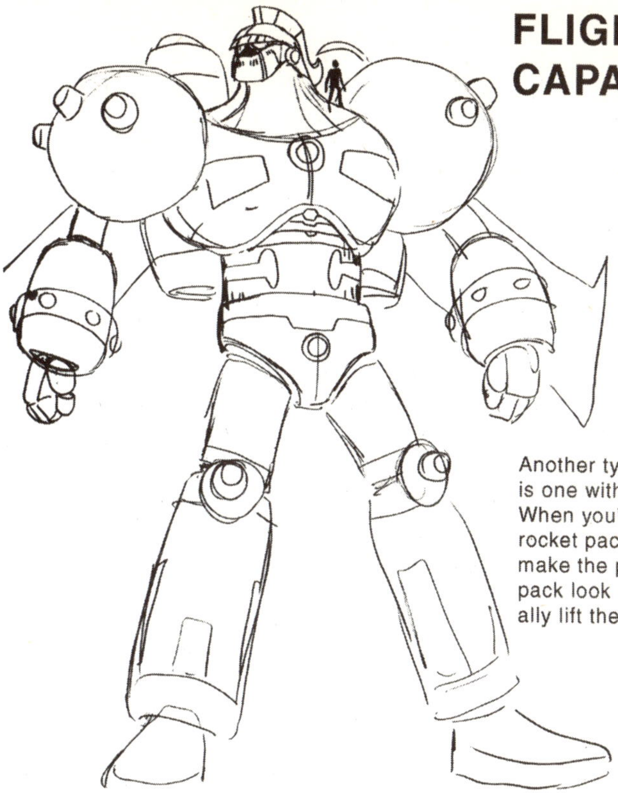

FLIGHT CAPABILITIES

Another type of heroic mech is one with flight capabilities. When you're designing a rocket pack for your mech, make the proportions of the pack look as if it could actually lift the mass of the mech.

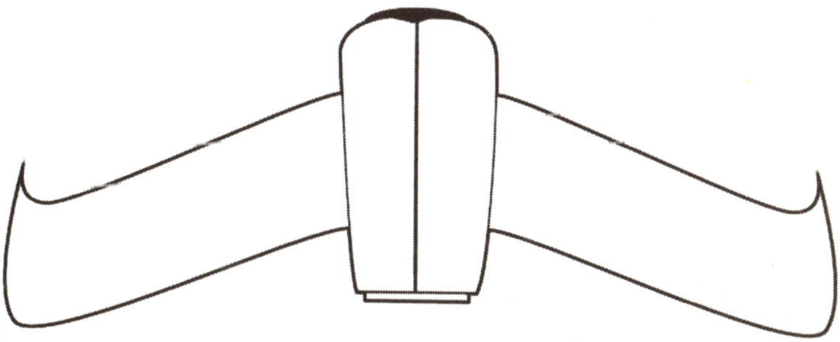

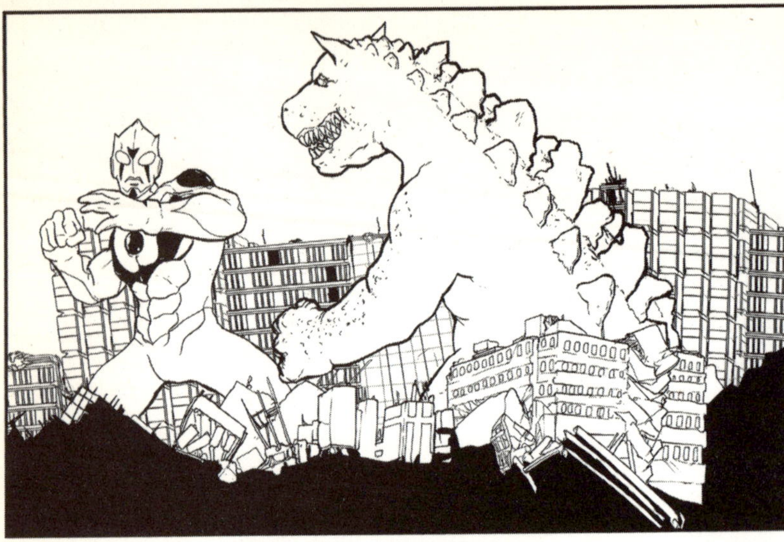

"MAN IN SUIT"

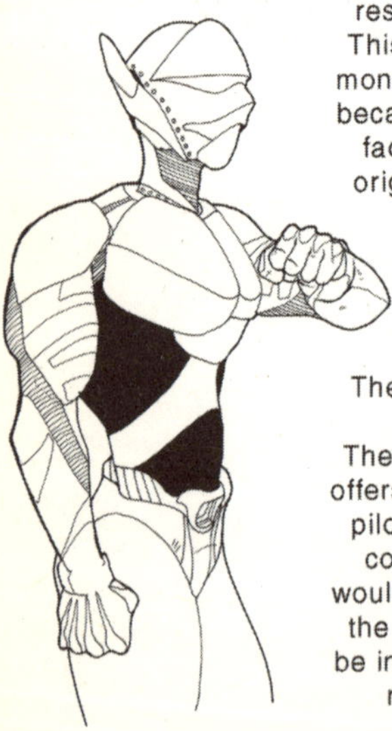

Sometimes heroic mecha can resemble humans in suits. This type of design is commonly used for heroic mecha because the villains the hero faces usually are alien in origin and the human form quickly helps associate the viewer with who the hero is.

The actual hero is the man inside the mecha. The cockpit inside the chest offers better protection for the pilot compared to the head cockpit, where the armor would be typically thinner and the head of the mech could be in danger of being disconnected from the body.

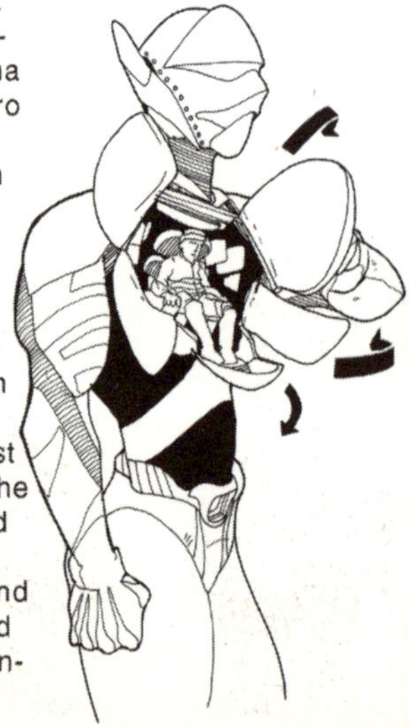

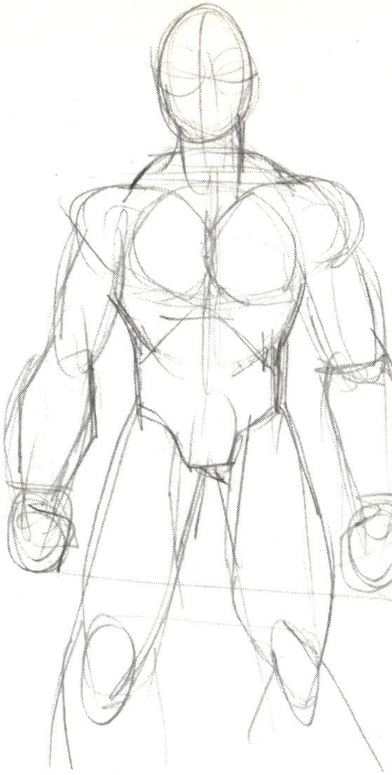

Many of the heroic mechs you will see have a humanoid configuration. As you see here, the shape of the mech starts like any other figure, using circles and ovals to lay out the figure.

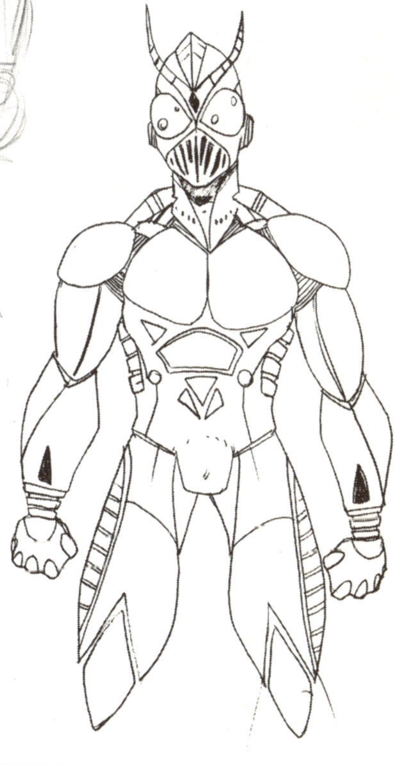

The pilot can sit in either the head or the chest.

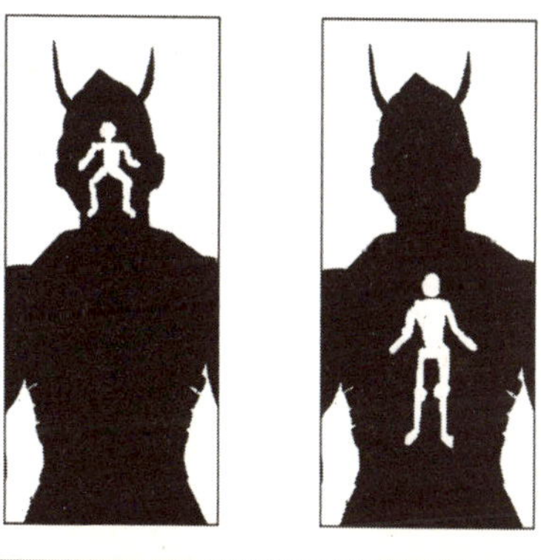

ADVERSARIES

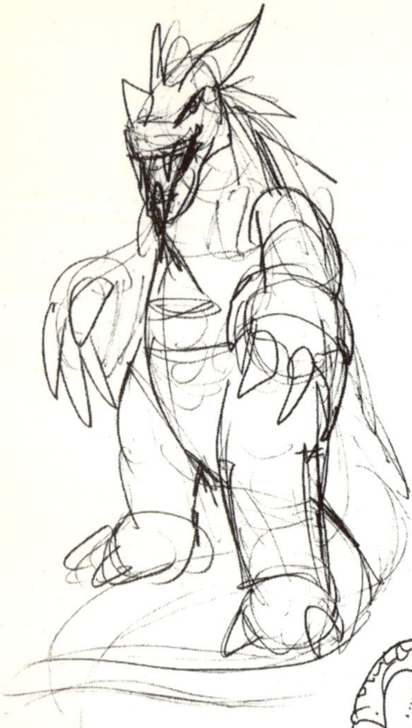

When the enemies are alien monsters, it's easier for the viewer to associate with the robotic hero.

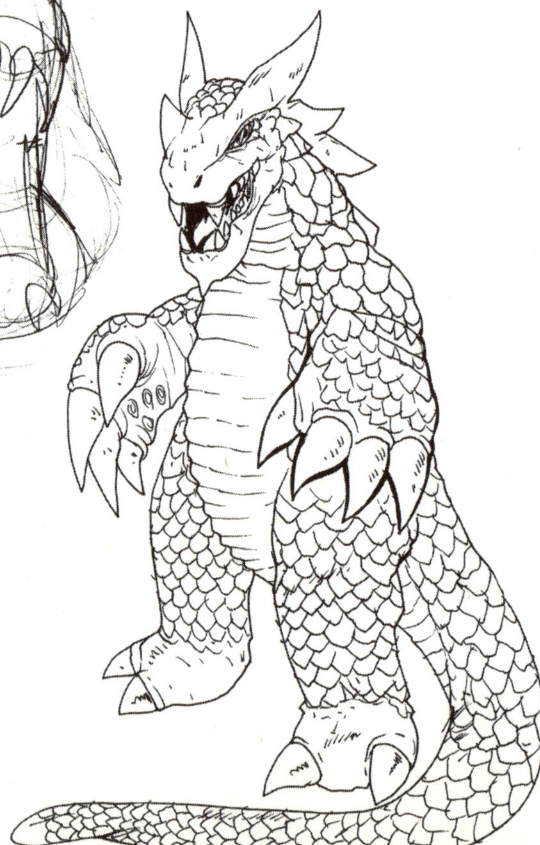

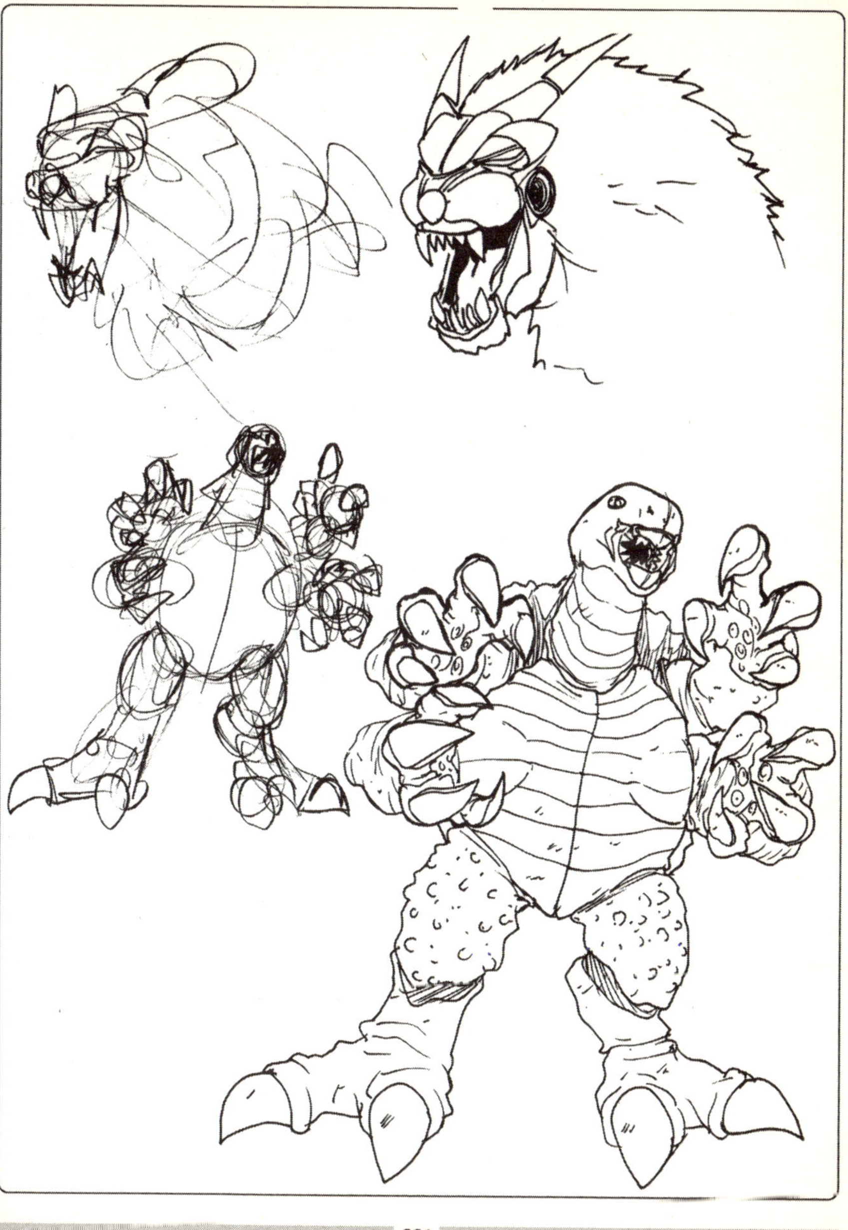

PULP-ORIENTED MECHA

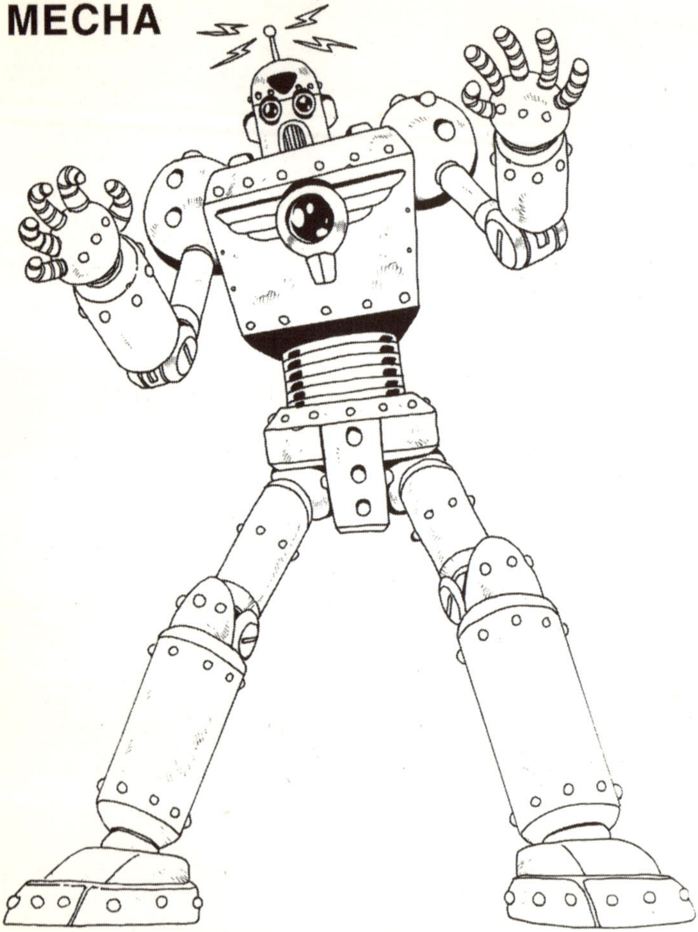

Mecha designs from the `30s and `40s were less worried about accuracy and more concerned about presenting a menacing-looking robot.

Pulp-oriented mecha represented the fears of industry and technology.

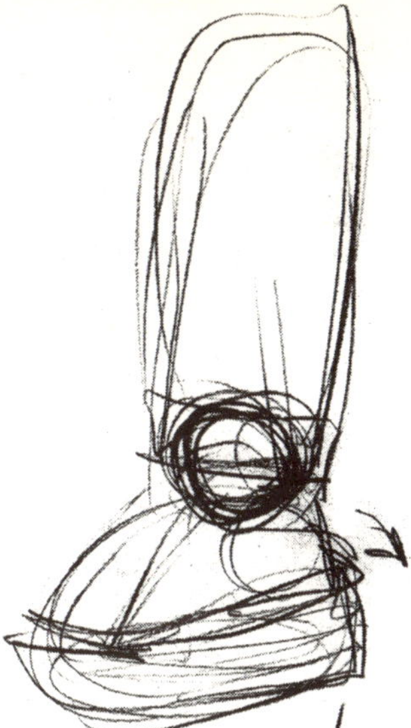

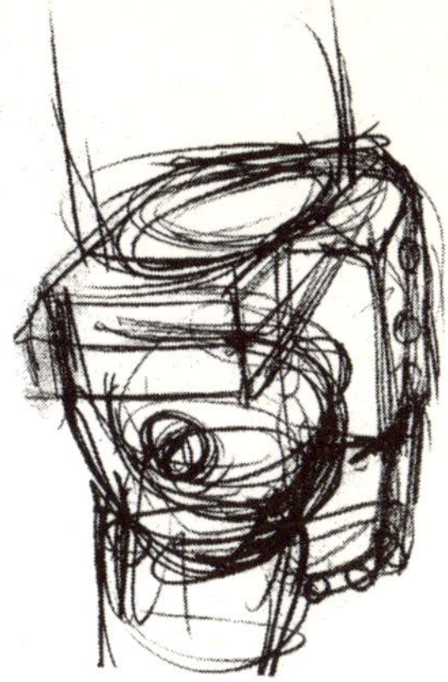

LEGS AND WAIST

The waist is made for simple turns and no flexibility. The bending takes place at the hips.
The feet are built heavy to support the robot's frame.

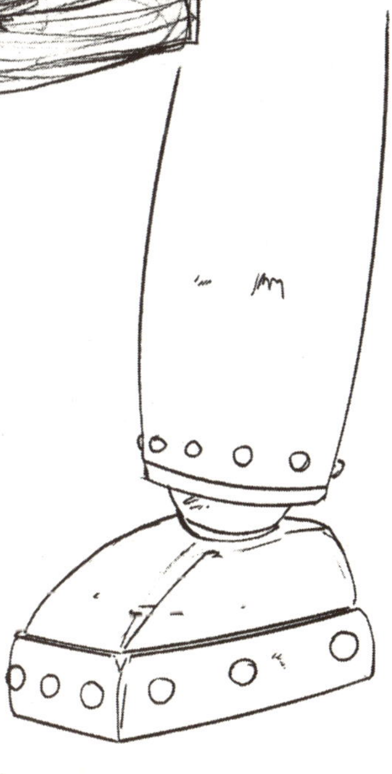

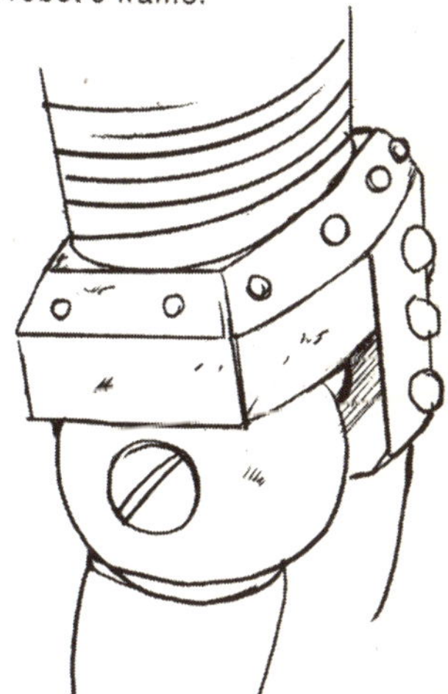

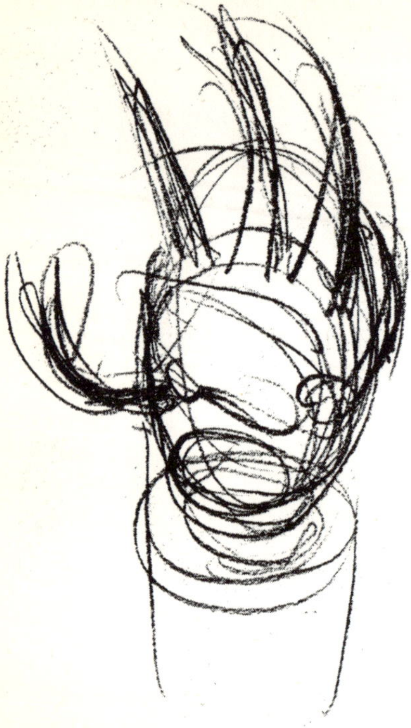

HANDS

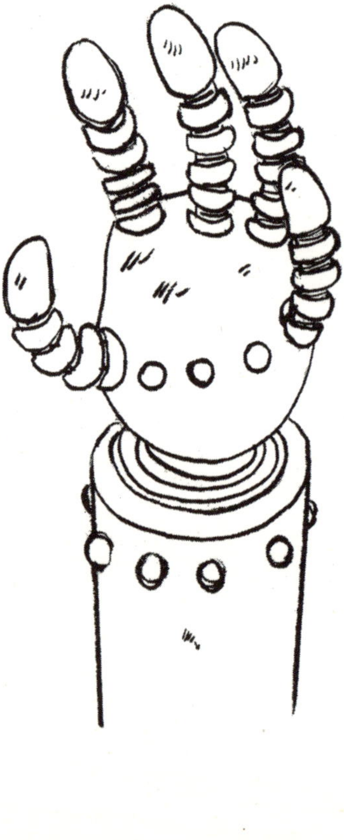

The hand itself is a solid, nonflexible mold. The three dots represent sensors. The fingers themselves do all of the gripping work. They're constructed like bendable tube wire.

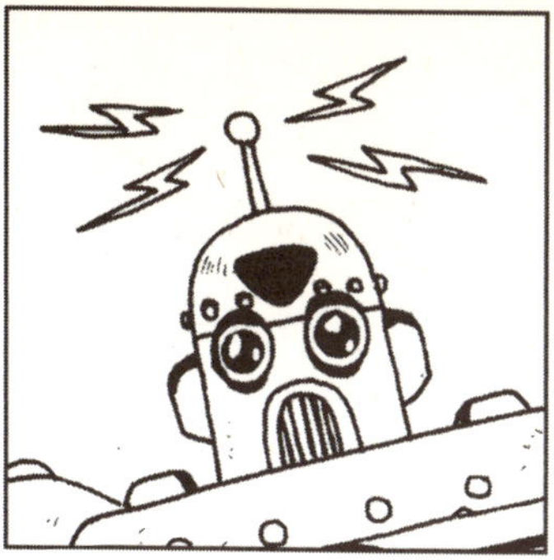

By referencing older-model radios and industrial designs for the face, we're able to create a unique-looking mech. With the rivets in the head, it looks welded together.

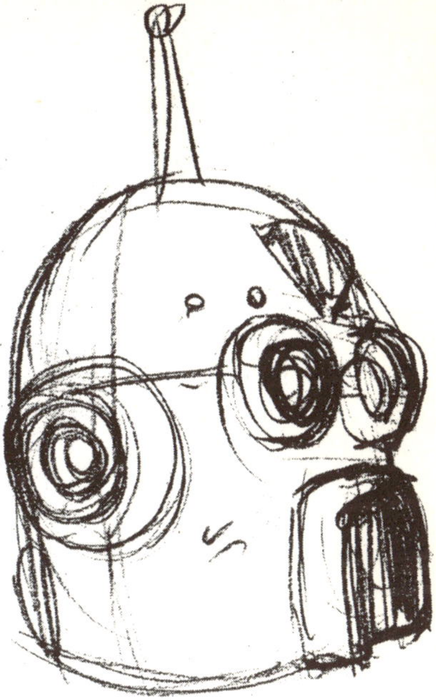

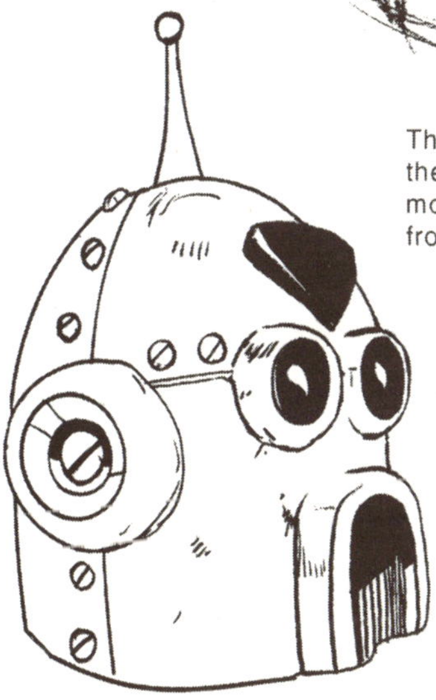

The mech has an antenna, so there's no need for ears. The mouth is represented by a grill from a `30s car.

ARM SETUP

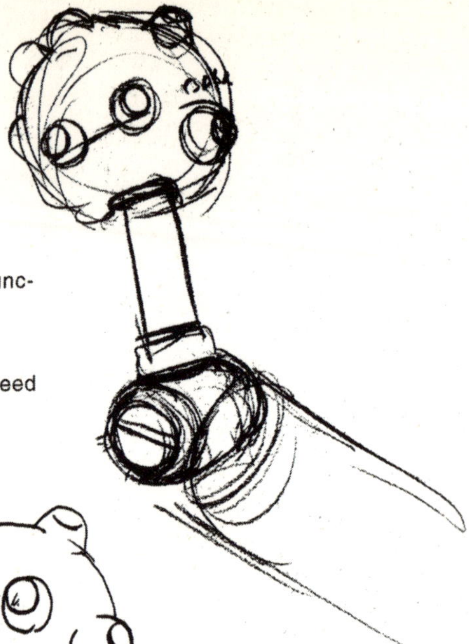

This arm contains the basic functions of mobility.

The arm works like a servo or pulley system, so there's no need for extra bulk.

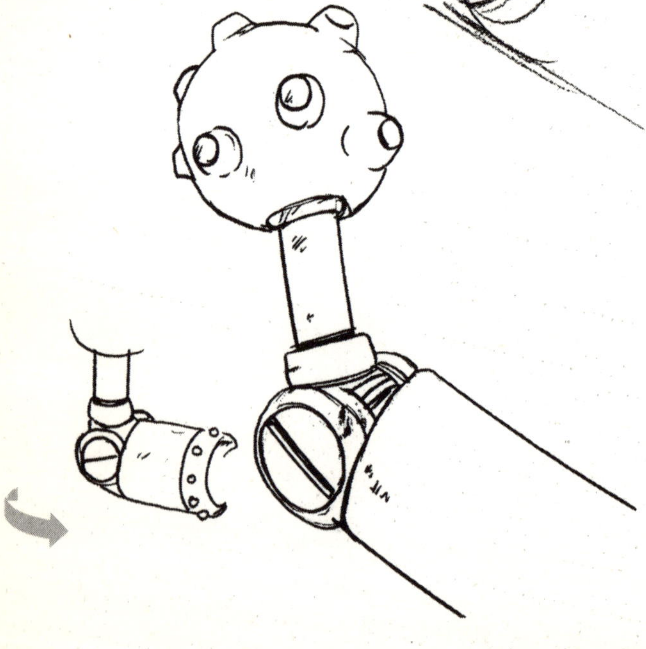

CHAPTER 06

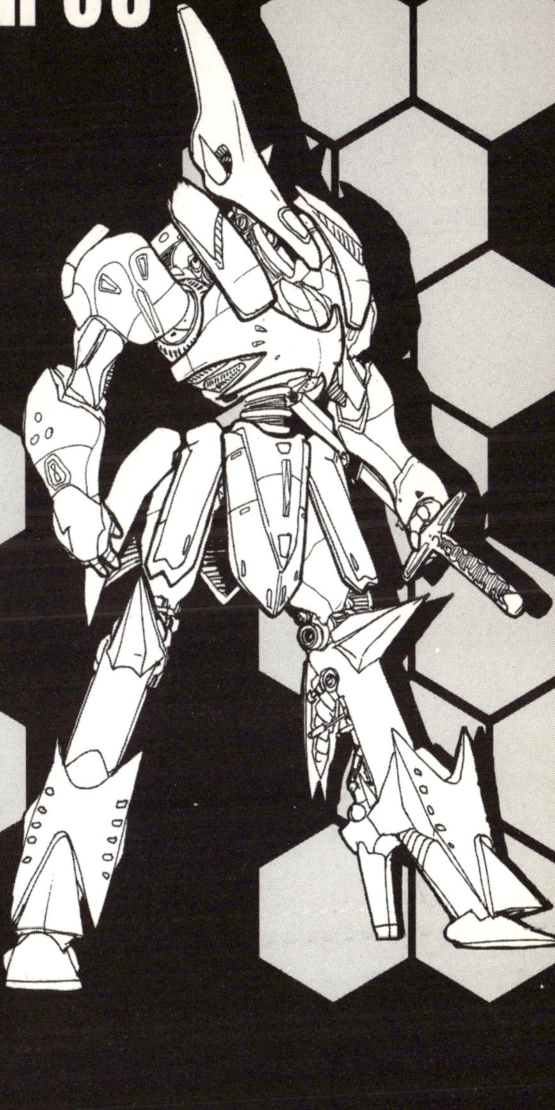

FANTASY MECHA

MEDIEVAL WEAPON ADD-ONS

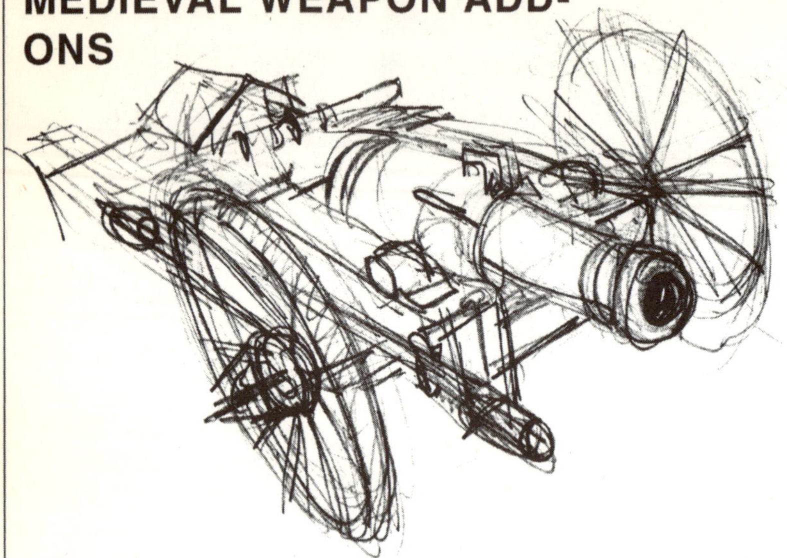

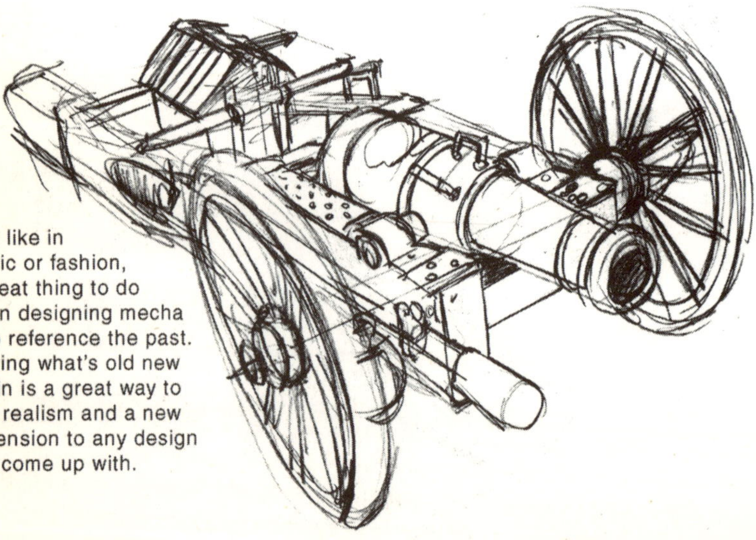

Just like in music or fashion, a great thing to do when designing mecha is to reference the past. Making what's old new again is a great way to add realism and a new dimension to any design you come up with.

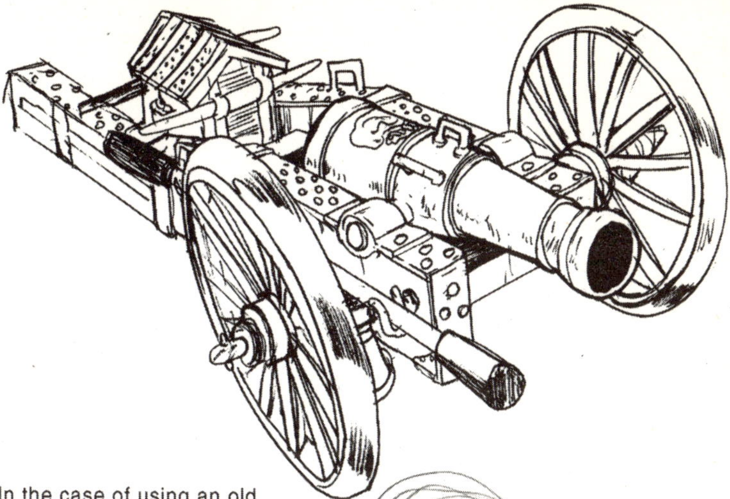

In the case of using an old cavalry cannon in a mech design, the overall shape lends itself to a forearm. By simply removing the wheels and changing the wood elements to a smoother metal look, we recreate a whole new mecha arm design.

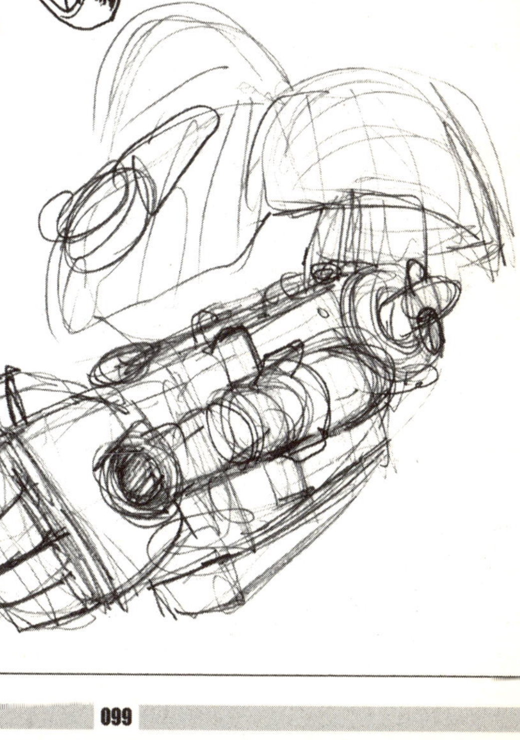

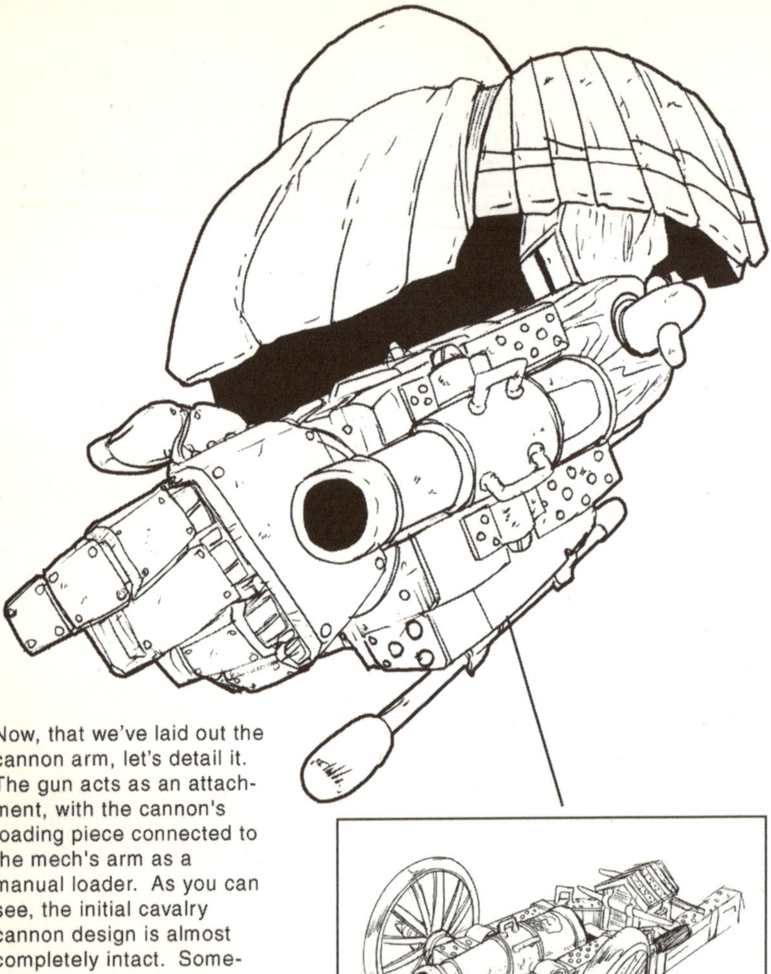

Now, that we've laid out the cannon arm, let's detail it. The gun acts as an attachment, with the cannon's loading piece connected to the mech's arm as a manual loader. As you can see, the initial cavalry cannon design is almost completely intact. Sometimes, not changing your source material too much can produce a great look.

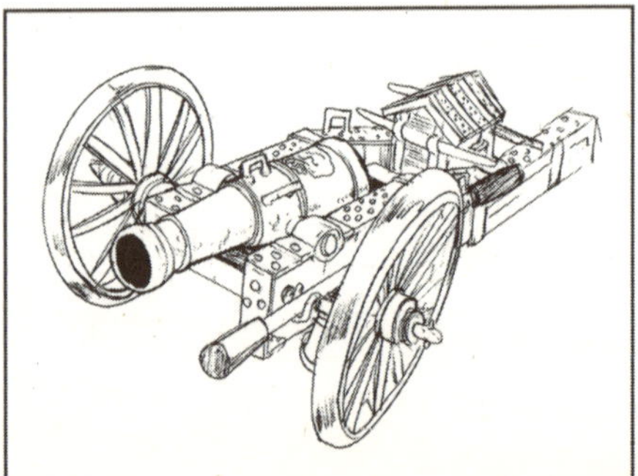

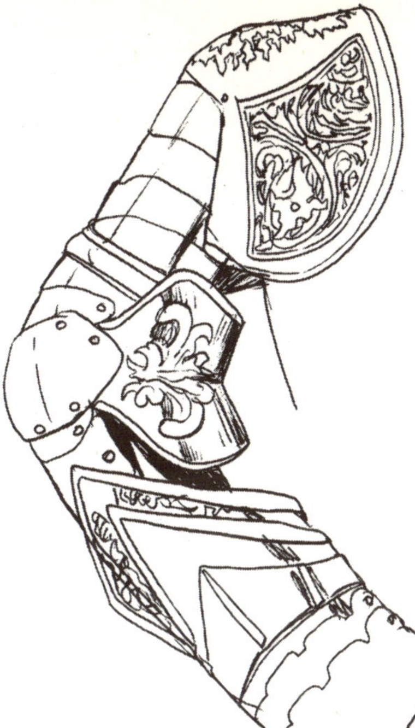

MEDIEVAL ARMOR

Let's go back a little farther in time to discuss armor. And what better place for armor than medieval times?
Suits of armor back then served one purpose: protect the wearers from sharp things that could hurt them. In designing mecha, the same principle applies, except in the case of the mecha, it's to protect sensitive electronics, hydraulics and joints.

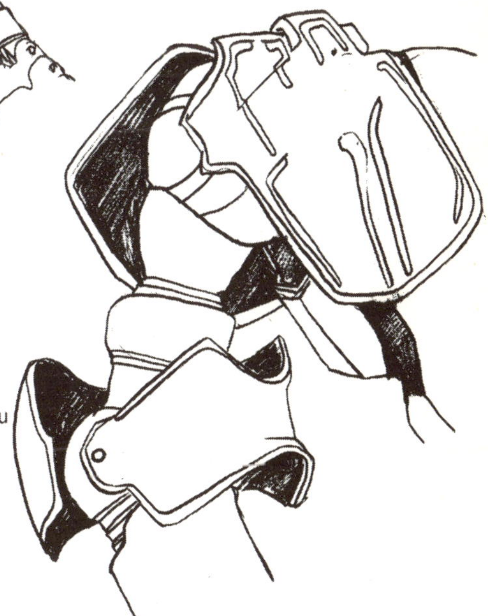

Suits of armor are made with portions that protect the joints. Try to emulate this in your design. Using actual suits of armor for reference will help you figure where to put certain armor pieces as well as how those pieces affect the body's movement.

HELMET DESIGN

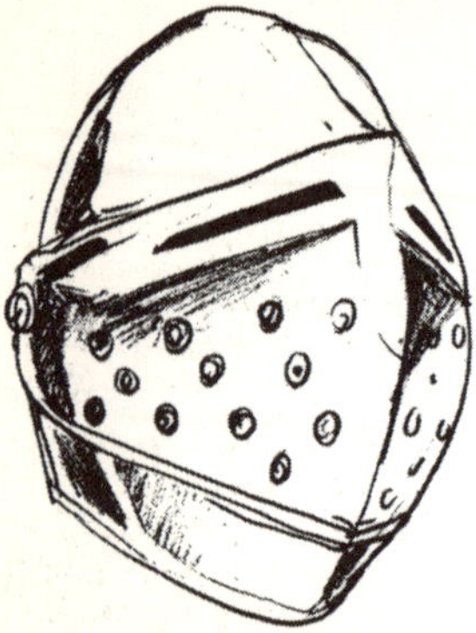

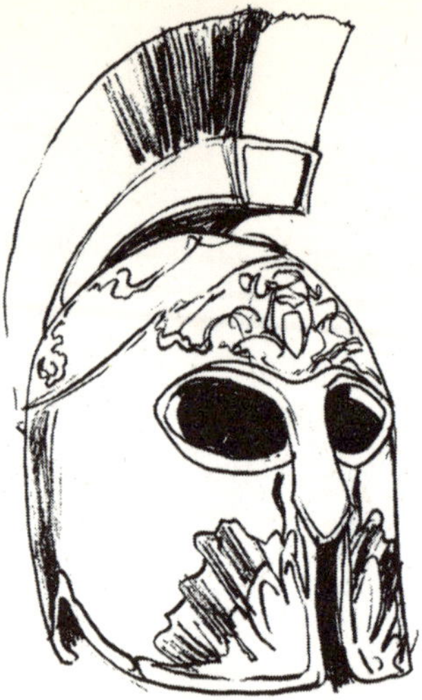

When you're choosing armor to reference, don't just limit yourself to medieval times. There are plenty of different styles of armor from different places and eras. All you have to do is do a little research. Classic helmets can be used as mech heads. This helps carry over the fantasy theme of your designs. On the next page, we have some examples of different styles.

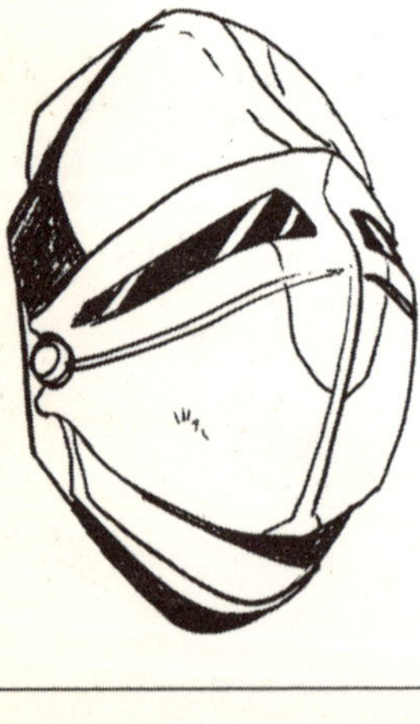

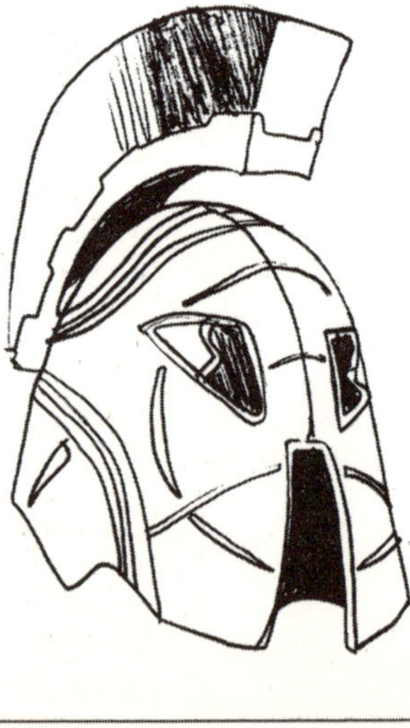

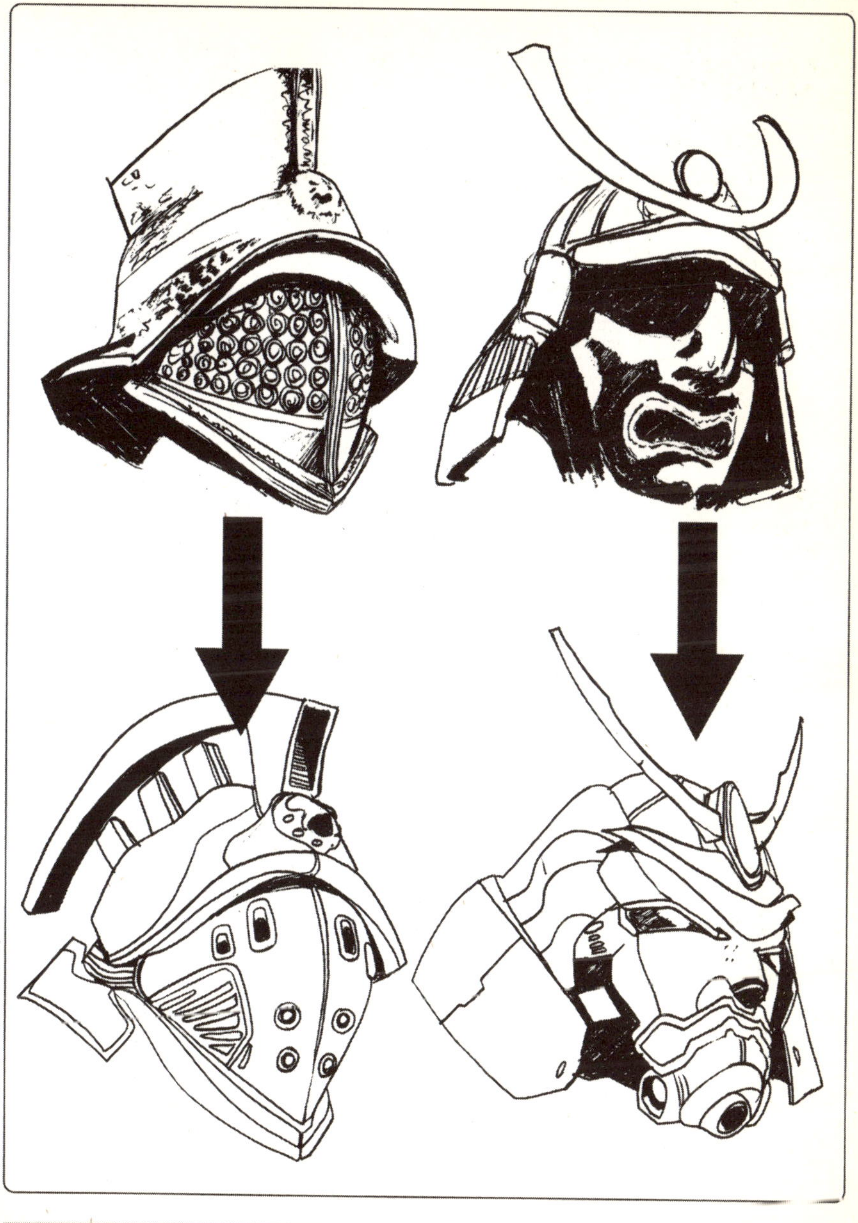

ANIMAL ARMOR

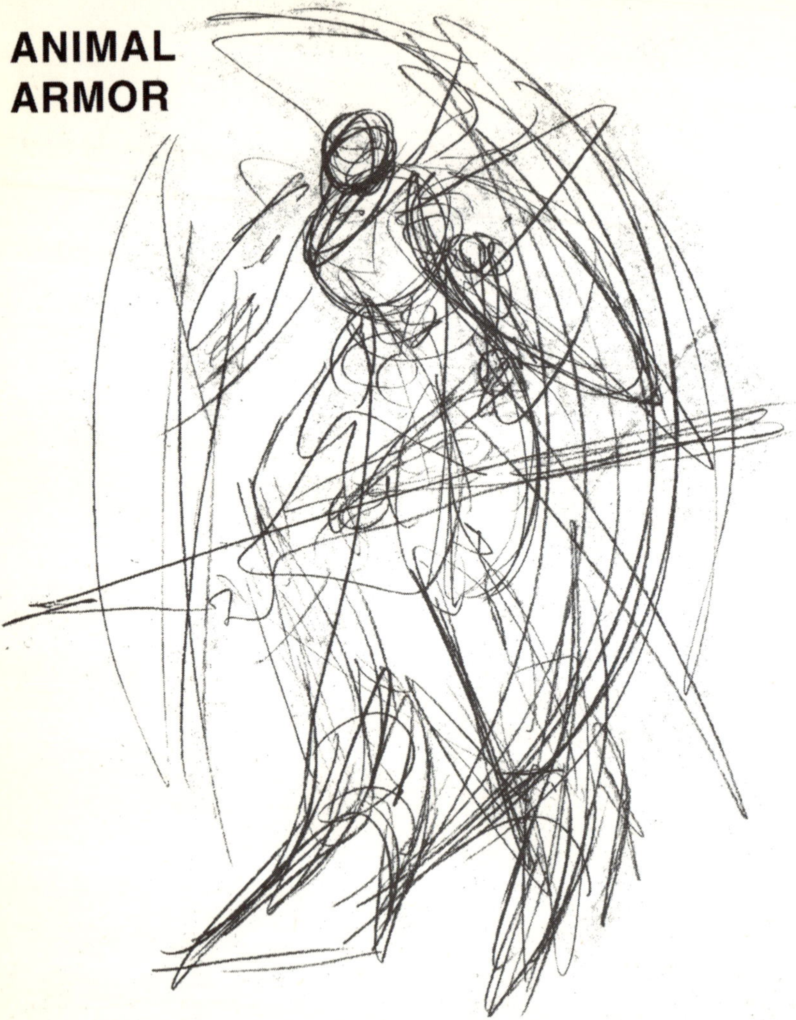

We're going to design a mech taking the animal design philosophy and the suit of armor approach. The form is organic. You can see the beginnings of a sword and shield standing out at this stage.

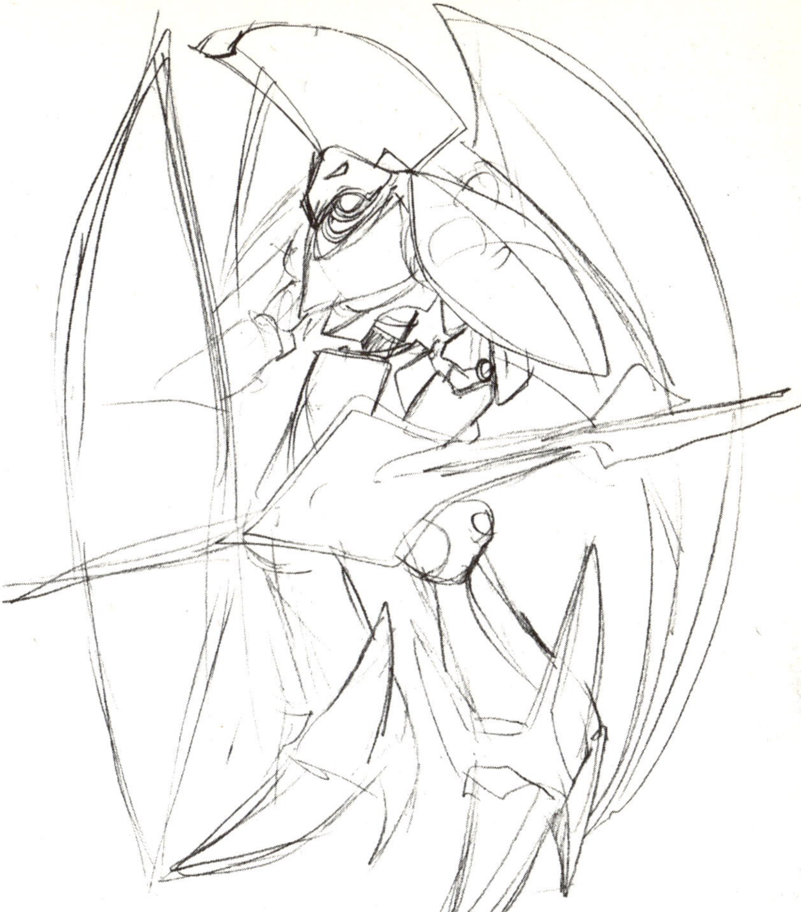

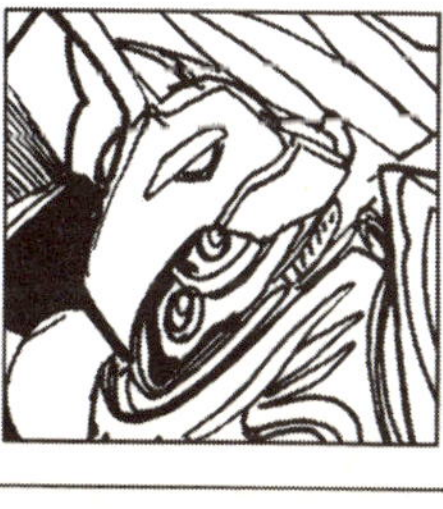

The overall look of this mech gives the impression of an eagle's face. The helmet has a feather-like mohawk. The wings on the back keep the theme consistent.

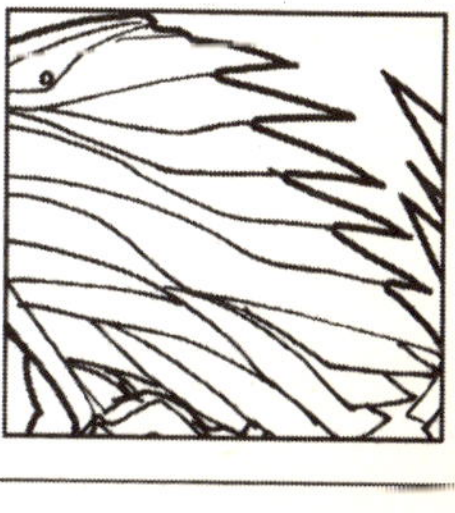

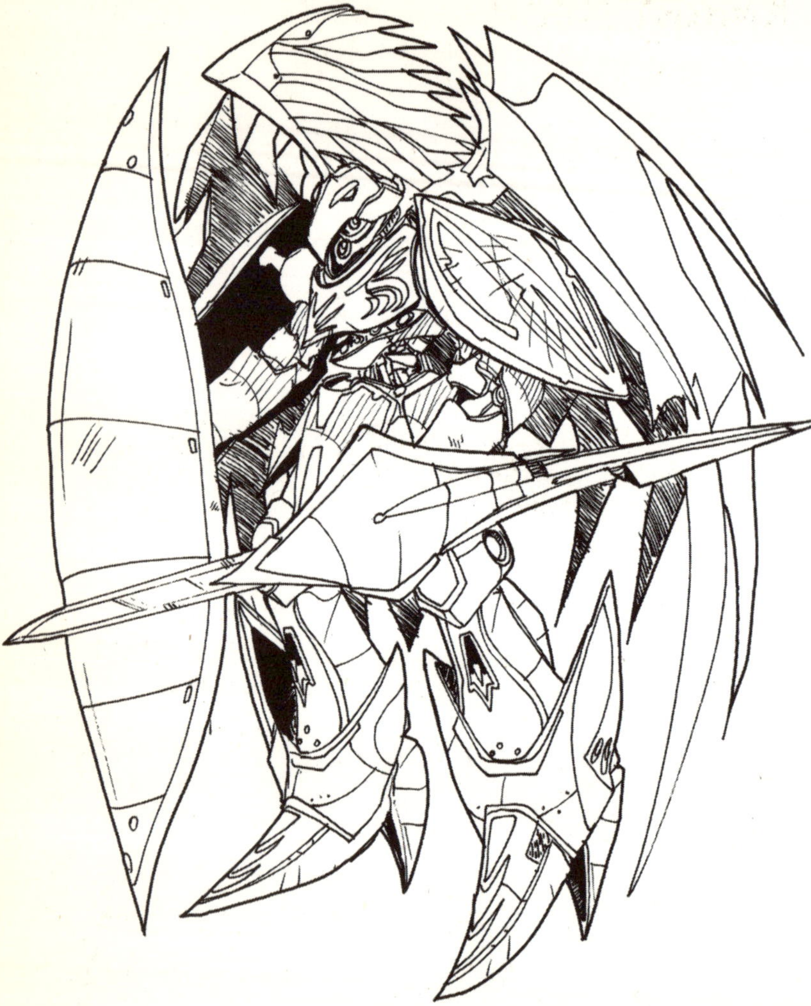

After you finish laying out the structure, it's time to finish the design. Details like plate bolts and markings help round out the drawing.

KNIGHT MECH

For this next illustration, we're going to work out a mech based on a medieval armor design. First we want to figure out the pose and lay out the structure.

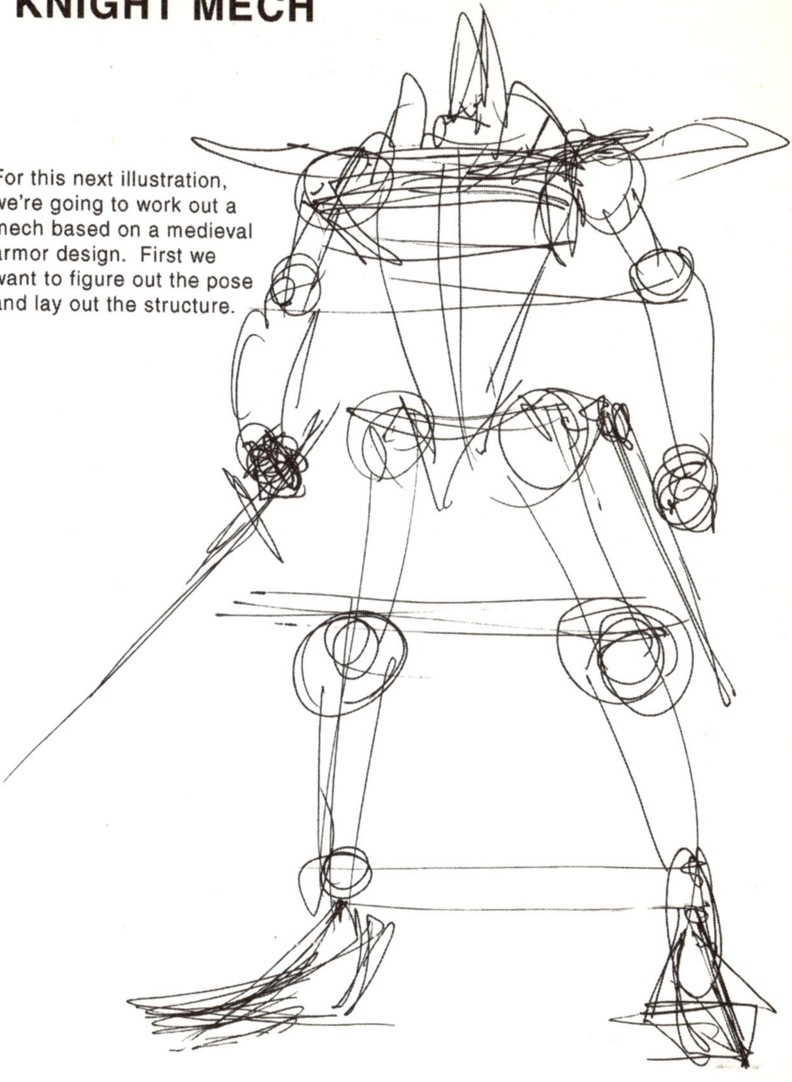

While you're working, remember that this suit of armor has to work without the benefit of a human inside. So when you're designing this mech, keep in mind that you have to figure out the underworkings like hips, shoulders, and elbows.

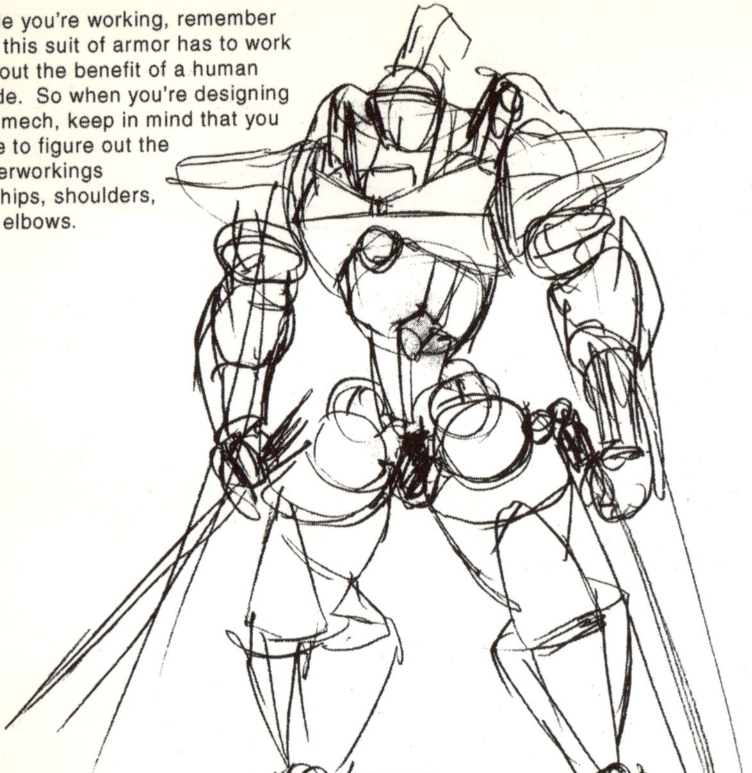

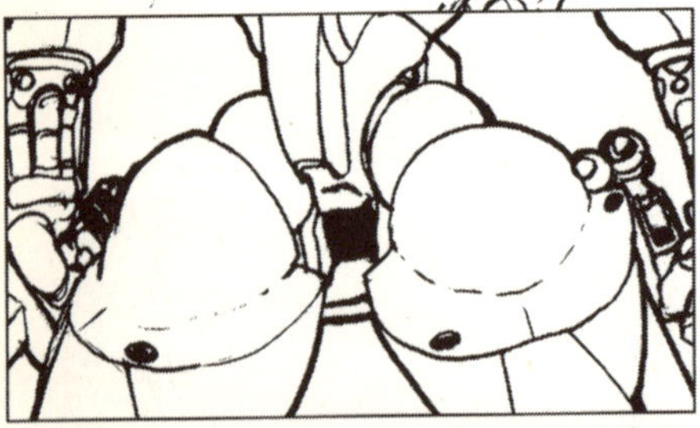

A close-up of the hip joints shows that they are ball-and-socket types. The armor is raised in the front of the exposed joints for added protection.

Here is the final drawing. As you can see, the places where the mech would be most vulnerable -- the knees, elbows, hips and shoulders -- now have additional armor. Take note of how the armor is cut and shaped so as not to restrict the movement of the mech.

Cheat
When designing new mechs, try blacking out the inner workings of the mech. It cuts down on your rendering time and it will help break up the whites on your character. But if it doesn't look good, DON'T DO IT!

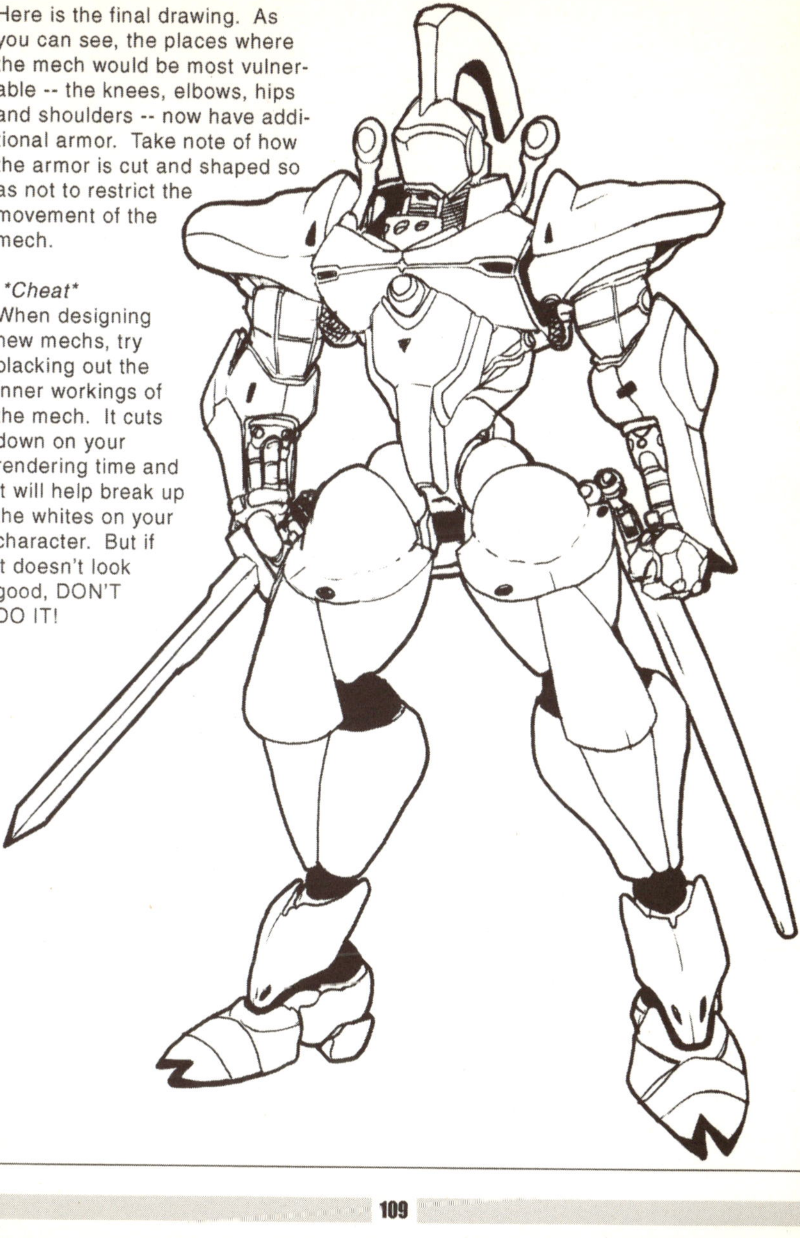

UNICORN MECH

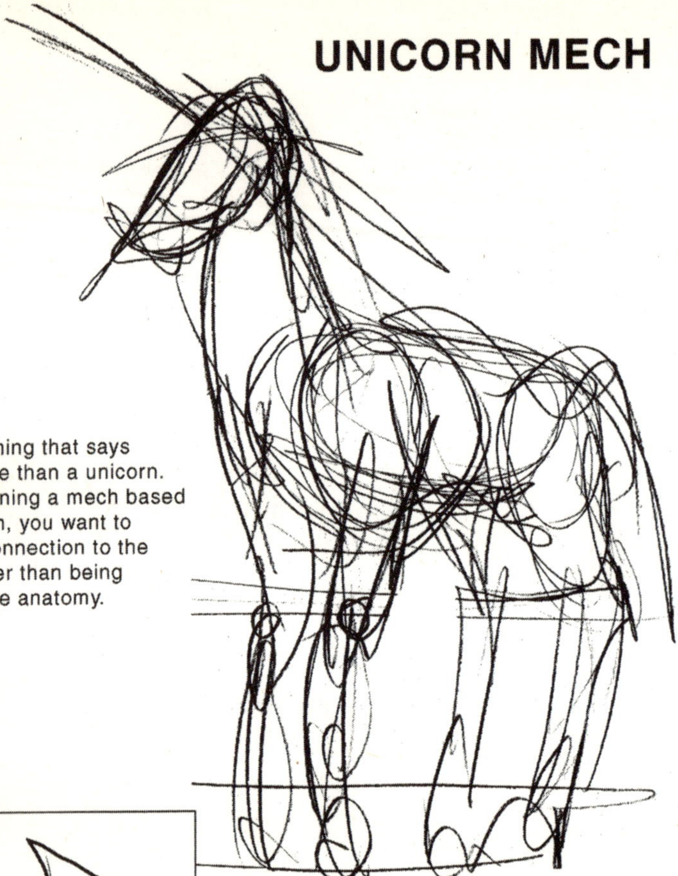

There's nothing that says fantasy more than a unicorn. When designing a mech based on a unicorn, you want to imply the connection to the animal rather than being bound by the anatomy.

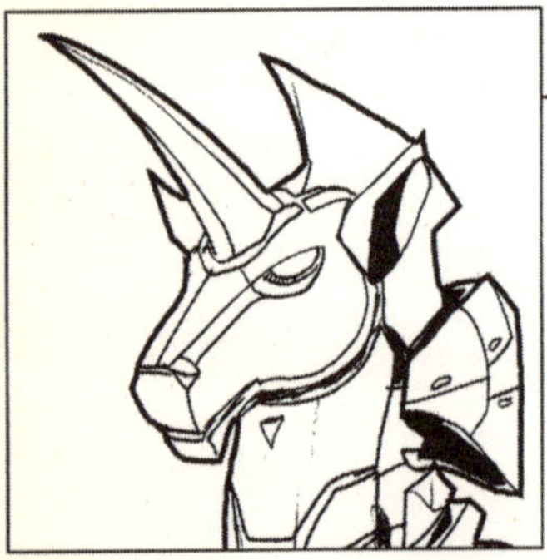

What you have to do is figure out what makes a unicorn a unicorn: the single horn and horse's body. Now when you're designing, the emphasis is placed on the horn and mane and how the entire body looks constructed. You can cheat on the anatomy a little to help it translate into a machine.

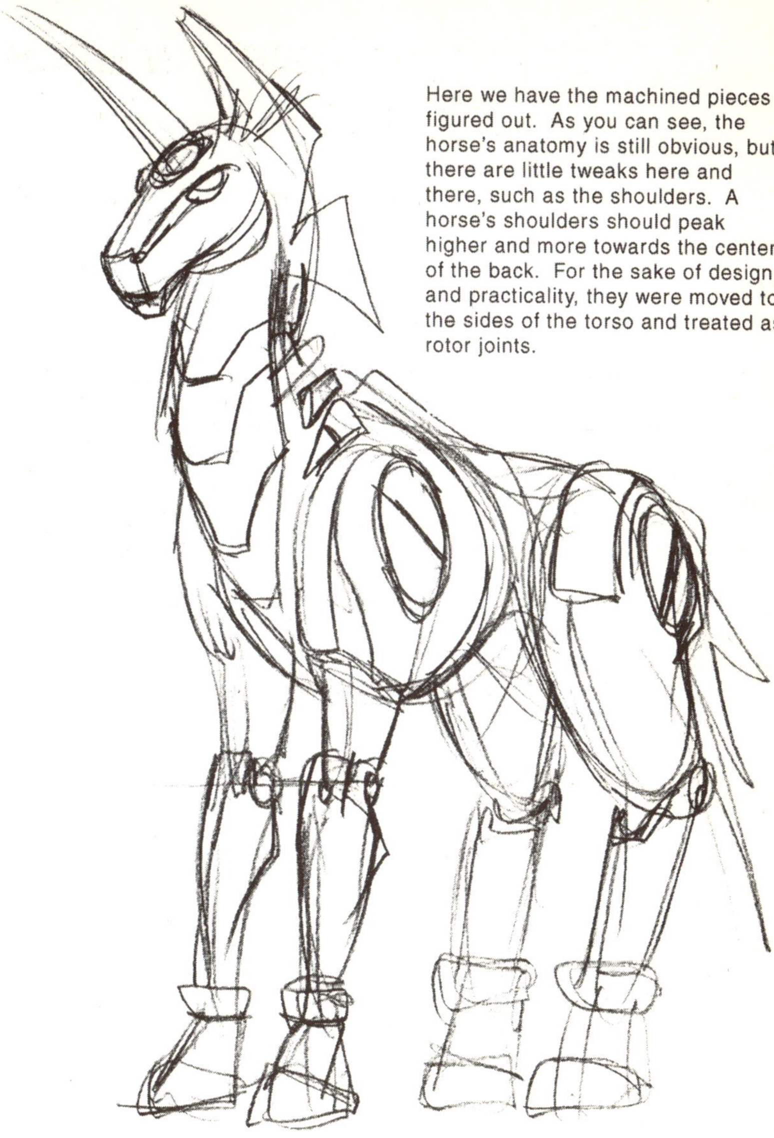

Here we have the machined pieces figured out. As you can see, the horse's anatomy is still obvious, but there are little tweaks here and there, such as the shoulders. A horse's shoulders should peak higher and more towards the center of the back. For the sake of design and practicality, they were moved to the sides of the torso and treated as rotor joints.

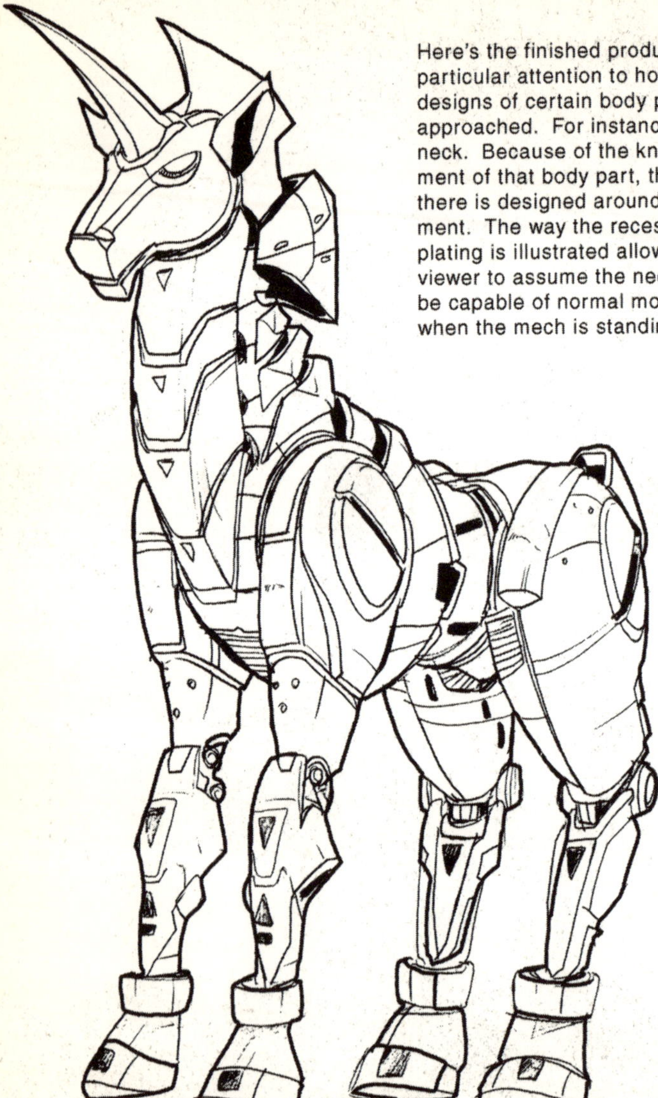

Here's the finished product! Pay particular attention to how the designs of certain body parts were approached. For instance, note the neck. Because of the known movement of that body part, the plating there is designed around the movement. The way the recessive plating is illustrated allows the viewer to assume the neck would be capable of normal motion, even when the mech is standing still.

CHAPTER 07

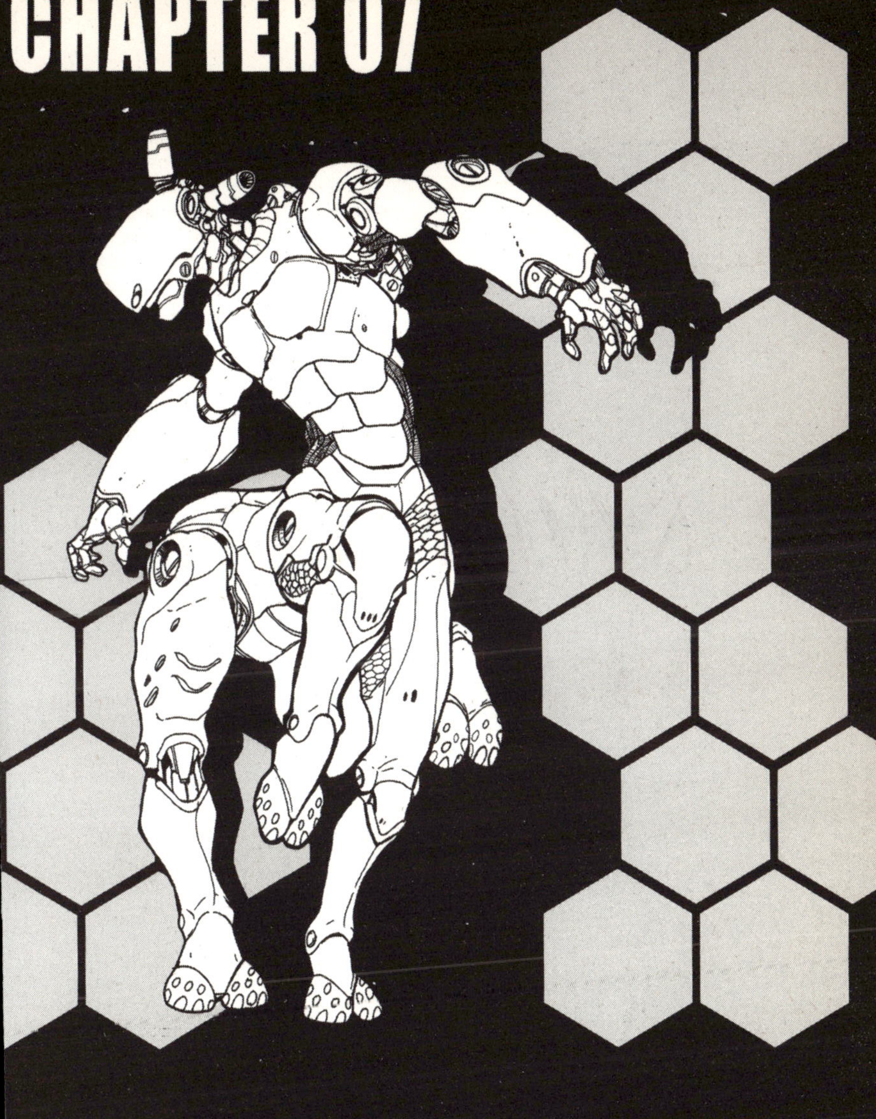

Animal Mecha

EFFICIENCY

Nature is efficient. All creatures are given exactly what they need to survive and propagate their species. Also, Nature knows not to waste space. Animals that weigh over a pound or are longer than 12 inches have between 0-4 legs, because any more than that wouldn't be necessary.
Also, everything in Nature is made to adapt to its environment.

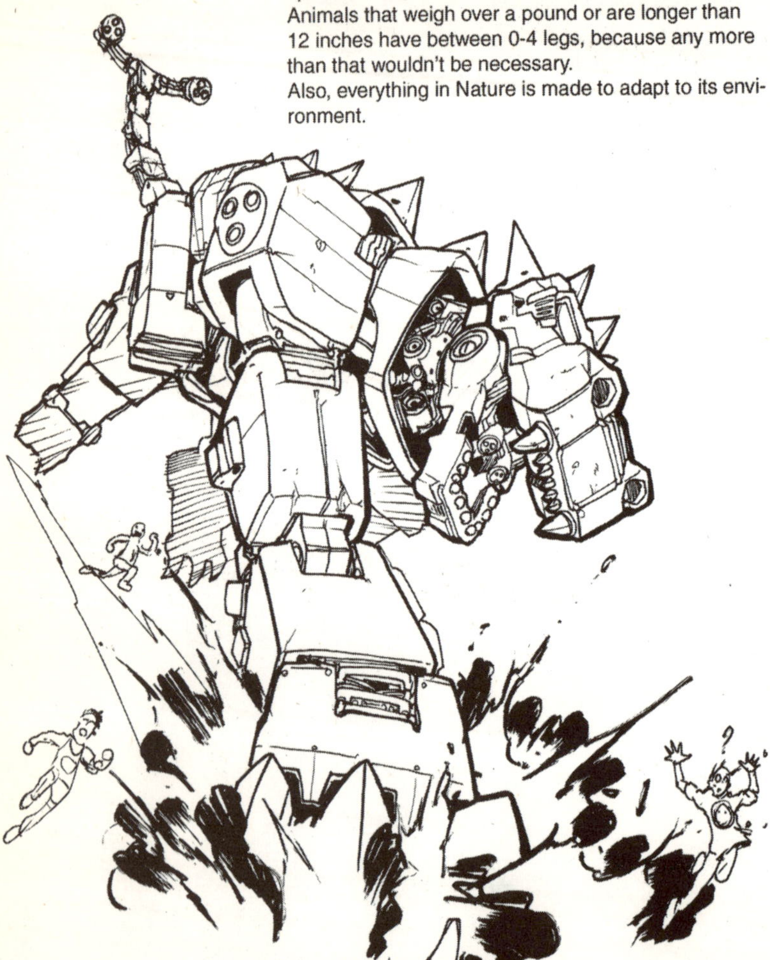

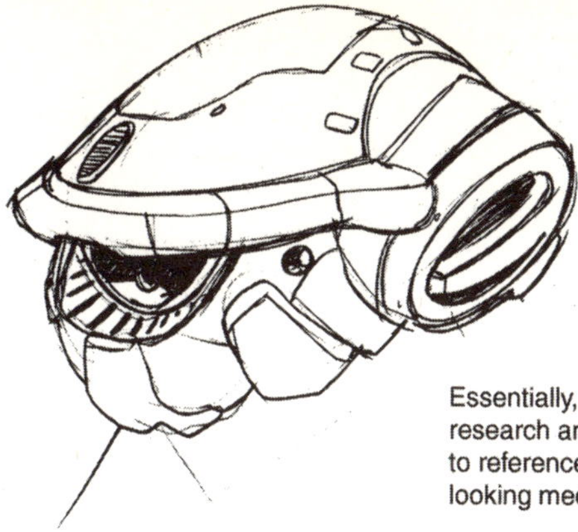

Essentially, we have millions of years of research and development right in front of us to reference when we want to create original-looking mecha.

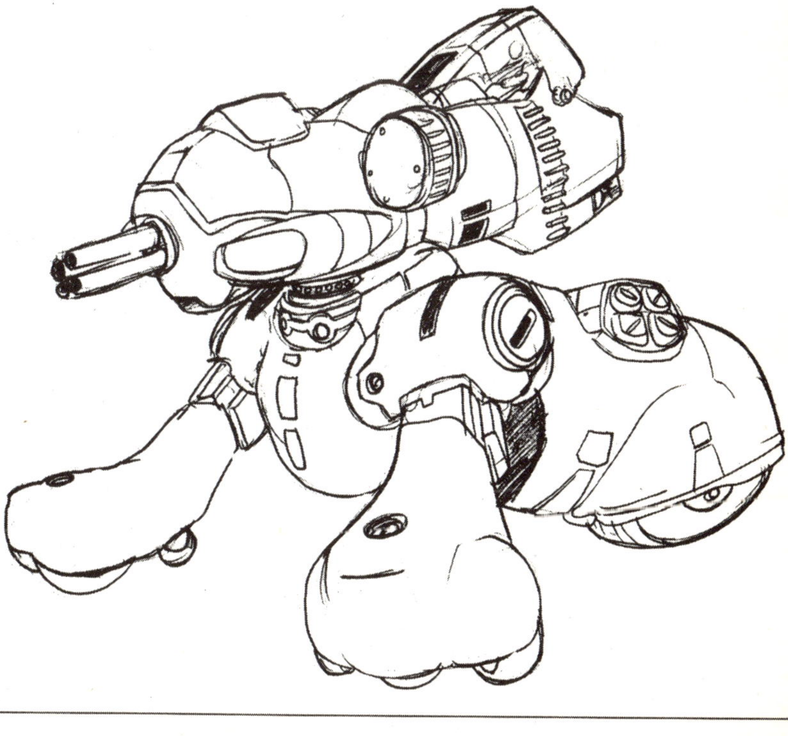

ANIMAL LAND MECHA

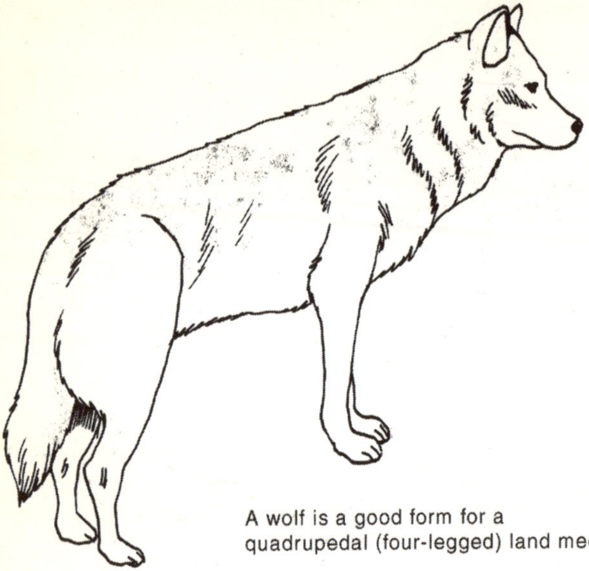

A wolf is a good form for a quadrupedal (four-legged) land mech.

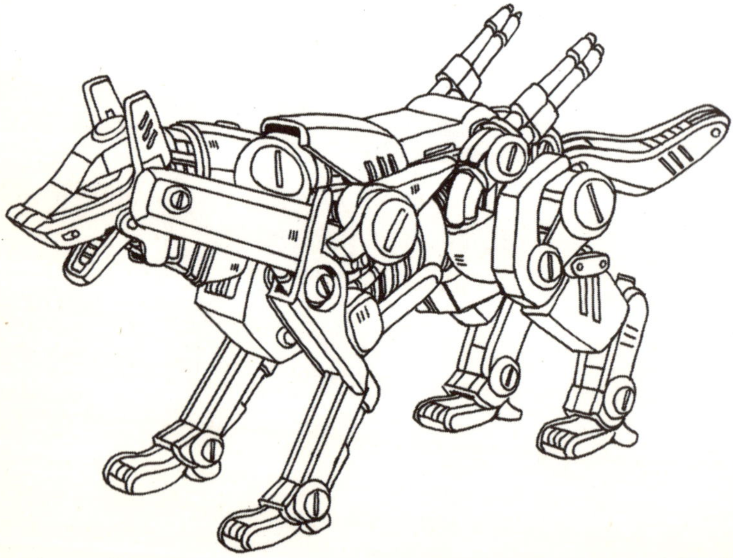

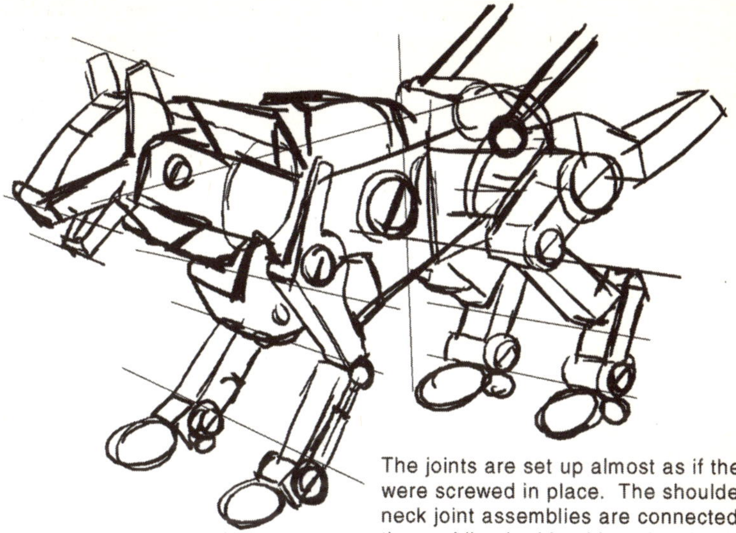

The joints are set up almost as if they were screwed in place. The shoulder and neck joint assemblies are connected, and the neckline is ridged into the shoulder.

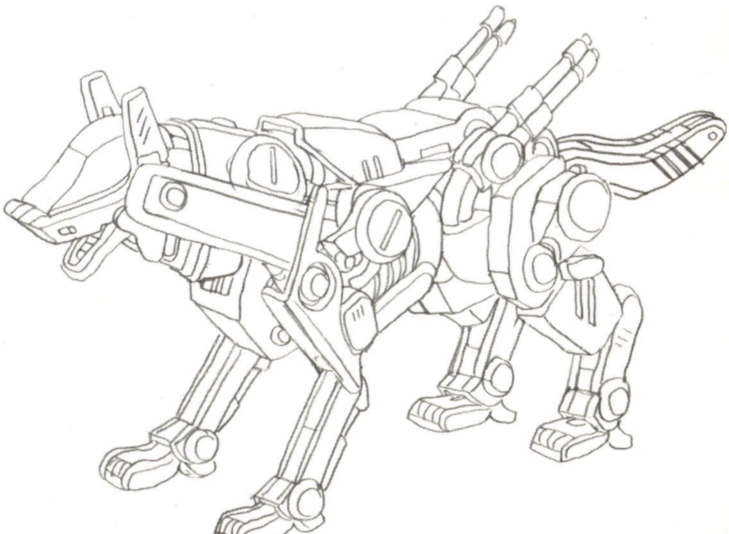

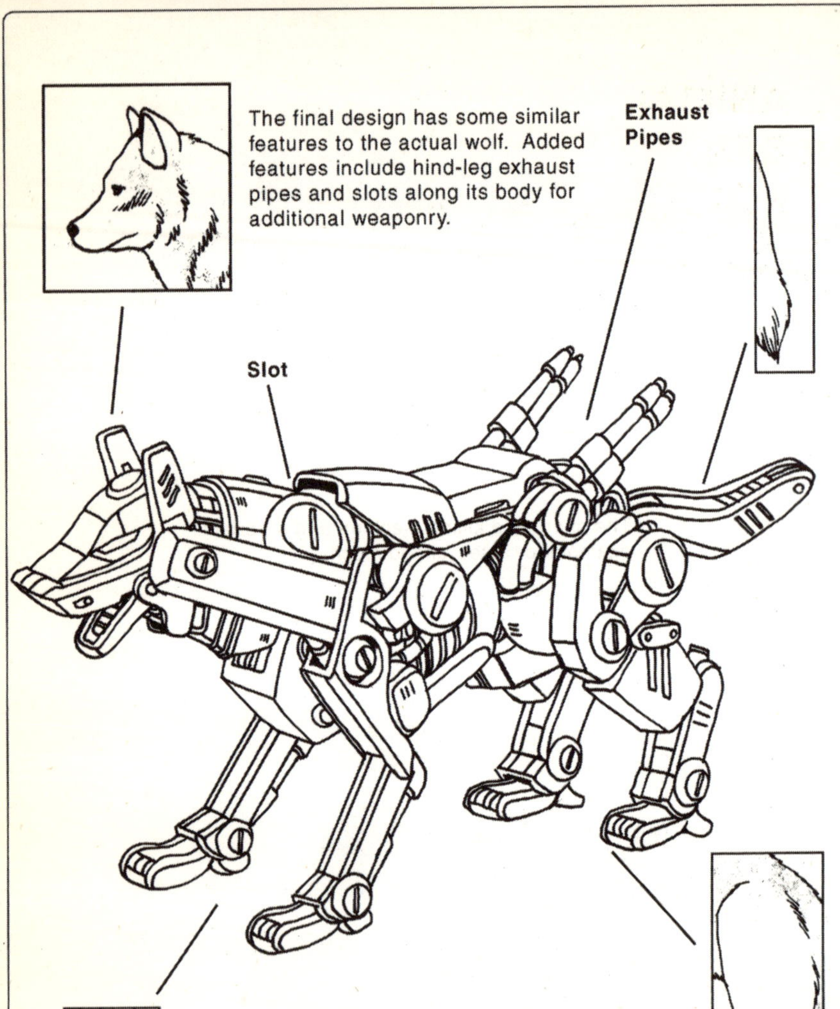
The final design has some similar features to the actual wolf. Added features include hind-leg exhaust pipes and slots along its body for additional weaponry.
Exhaust Pipes
Slot
The way the joints are grooved allows for the possible addition of armor and weaponry there.

ANIMAL SEA MECHA

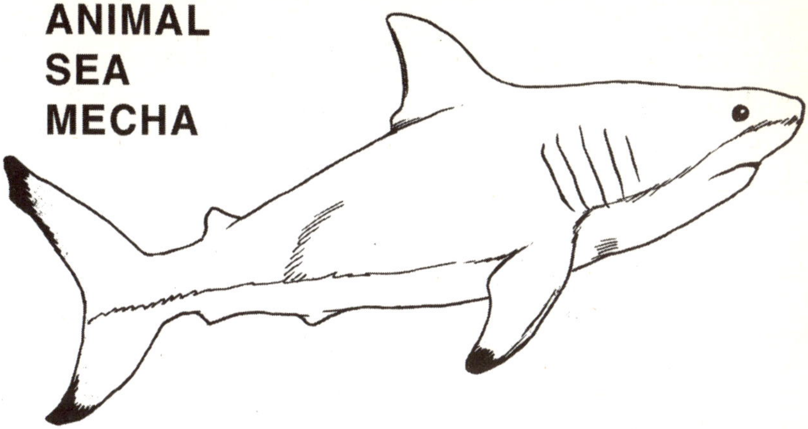

The shark is a killing machine and a perfect example for a sea-based mech.

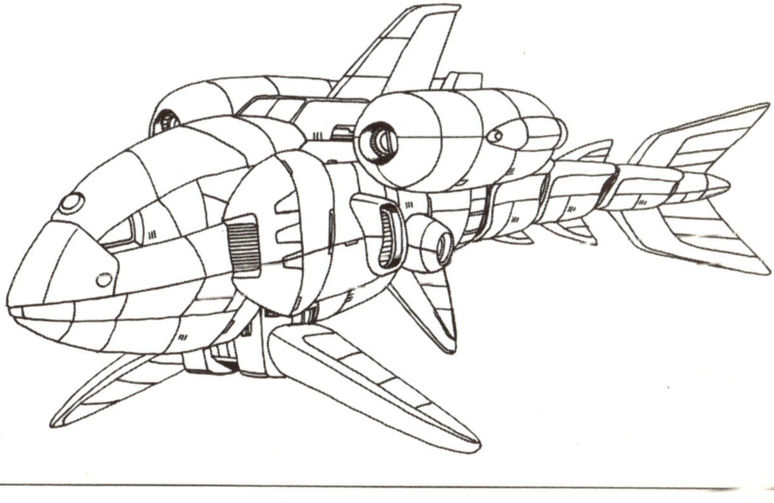

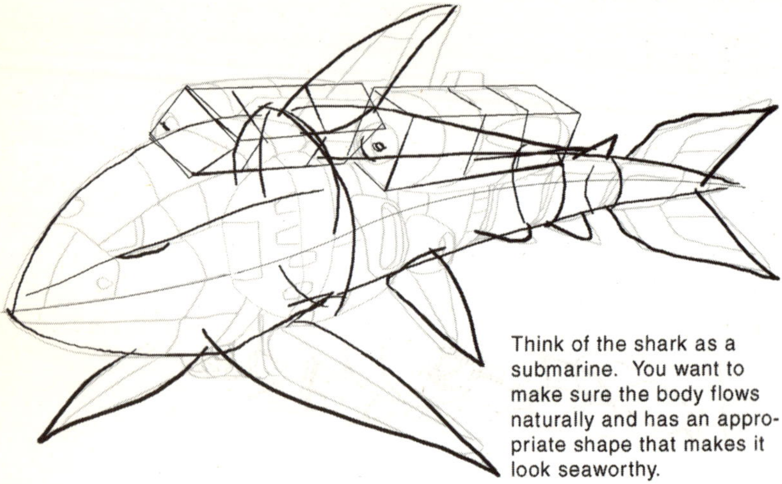

Think of the shark as a submarine. You want to make sure the body flows naturally and has an appropriate shape that makes it look seaworthy.

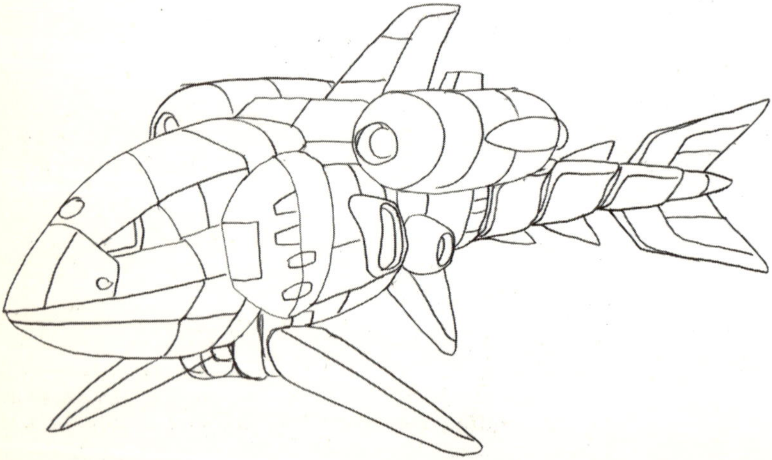

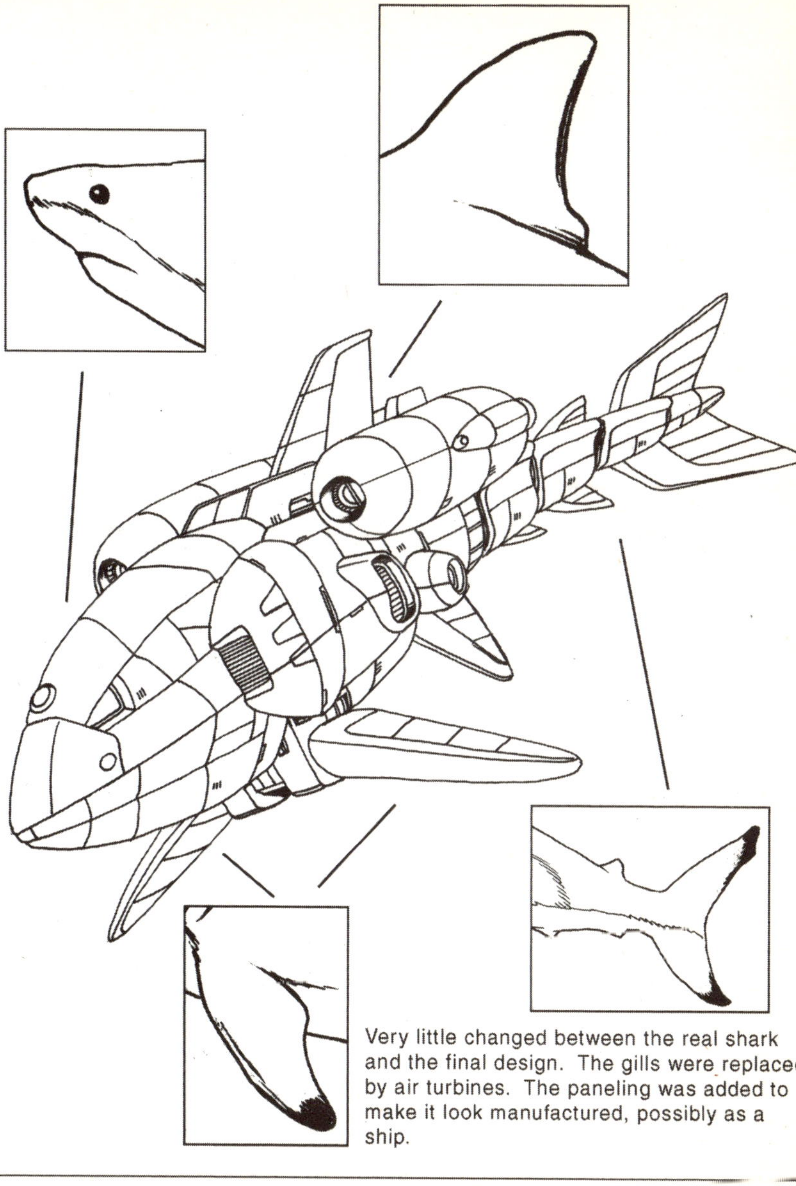

Very little changed between the real shark and the final design. The gills were replaced by air turbines. The paneling was added to make it look manufactured, possibly as a ship.

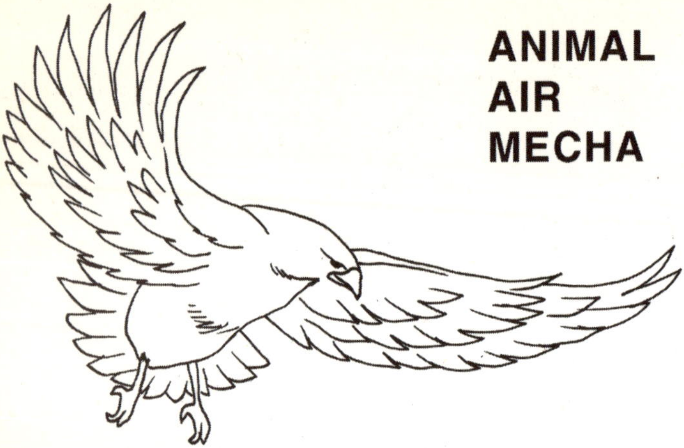

ANIMAL AIR MECHA

The hawk is an air-based predator and a good example for air-based mechs.

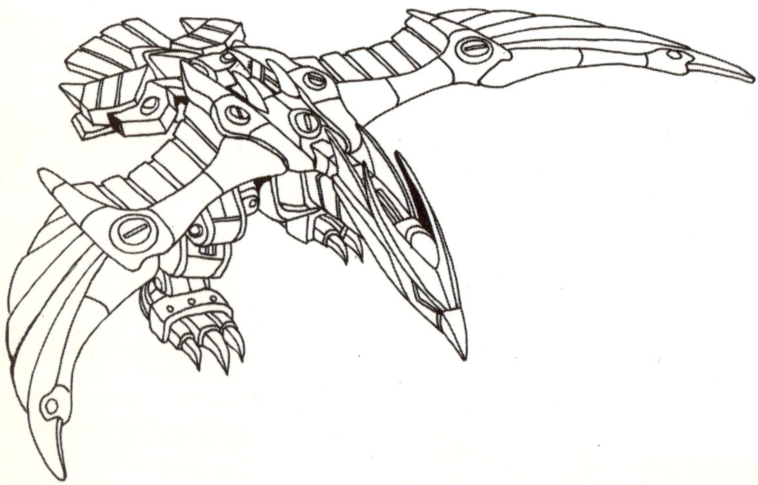

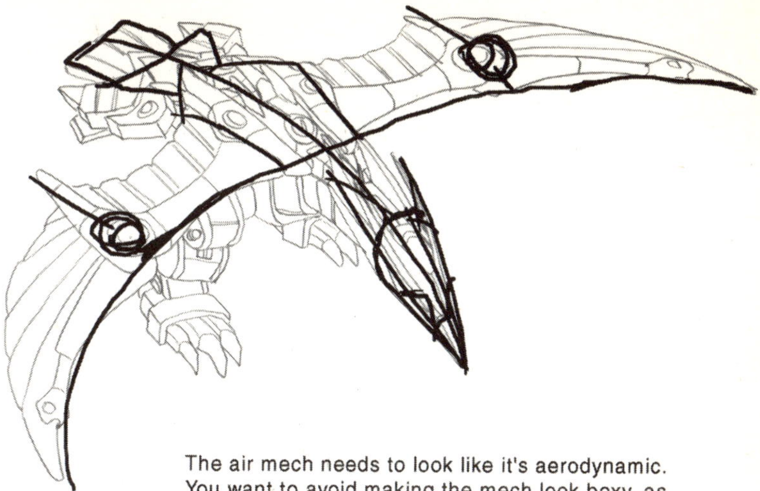

The air mech needs to look like it's aerodynamic. You want to avoid making the mech look boxy, as this doesn't give the impression it could fly.

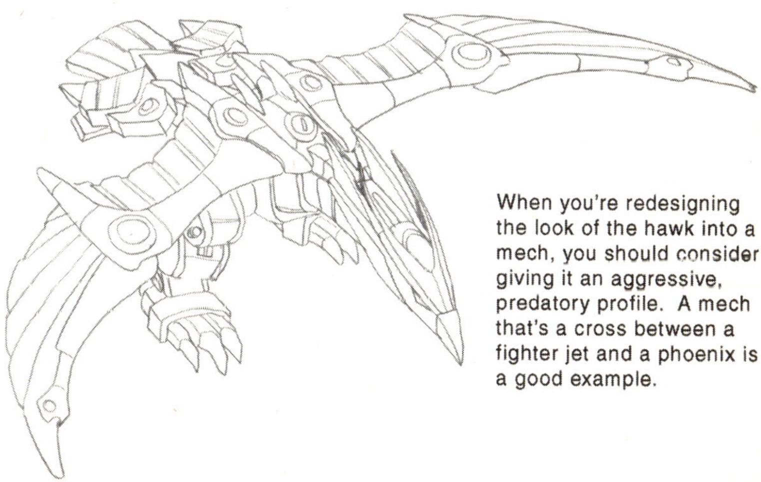

When you're redesigning the look of the hawk into a mech, you should consider giving it an aggressive, predatory profile. A mech that's a cross between a fighter jet and a phoenix is a good example.

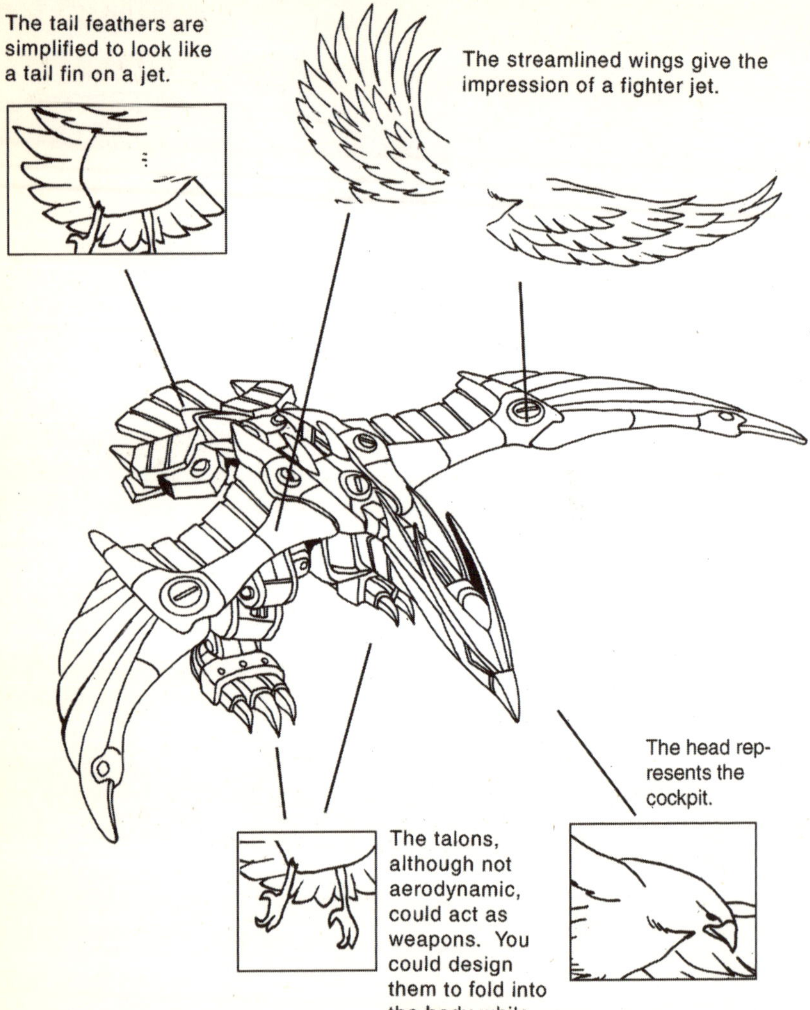
The tail feathers are simplified to look like a tail fin on a jet.
The streamlined wings give the impression of a fighter jet.
The head represents the cockpit.
The talons, although not aerodynamic, could act as weapons. You could design them to fold into the body while the mech is in flight.

ANIMAL GROUND MECHA

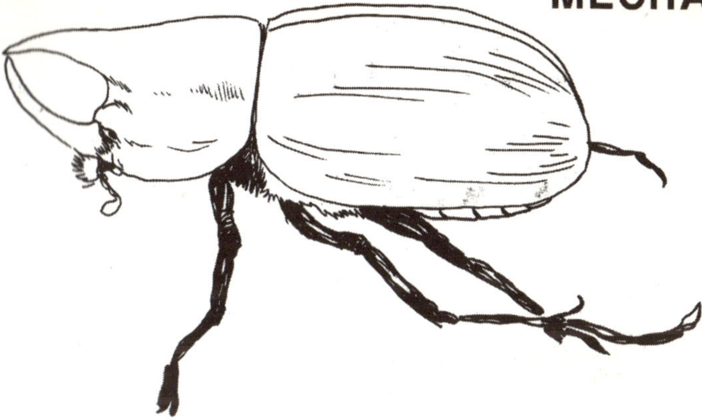

The beetle represents the land mech storm trooper, possibly for off-world combat. It's made to replicate not just the look, but also the behavior of an insect in a particular environment.

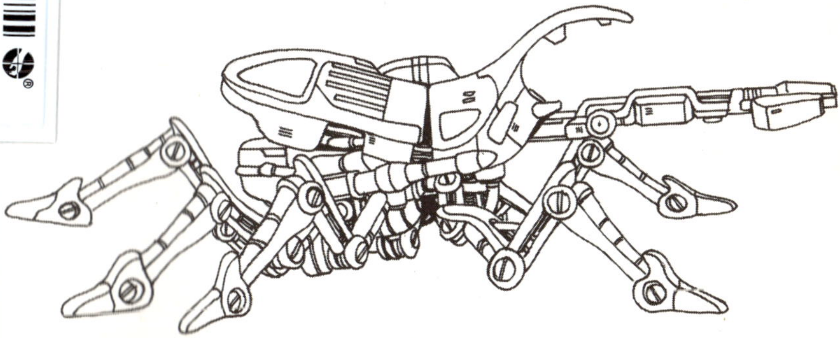

Depending on this mech's size, it could be used for surveillance or (if piloted) ground support.

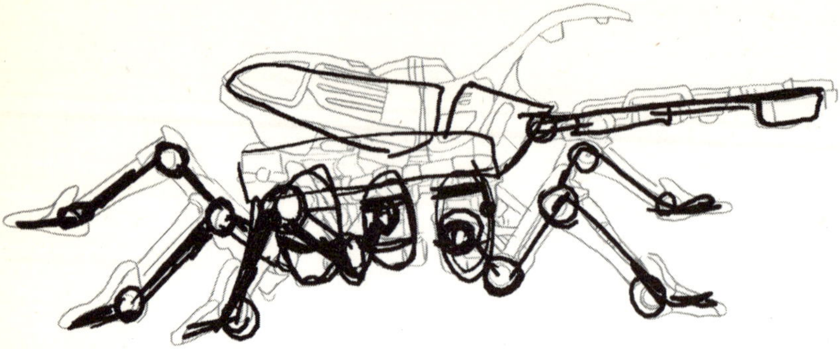

Insects' legs work in tandem. When designing a mech bug, try separating the legs into pairs.

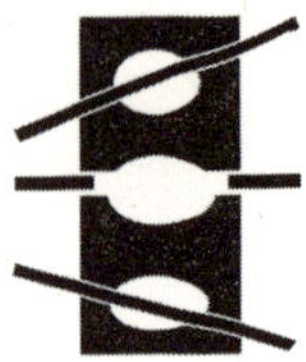

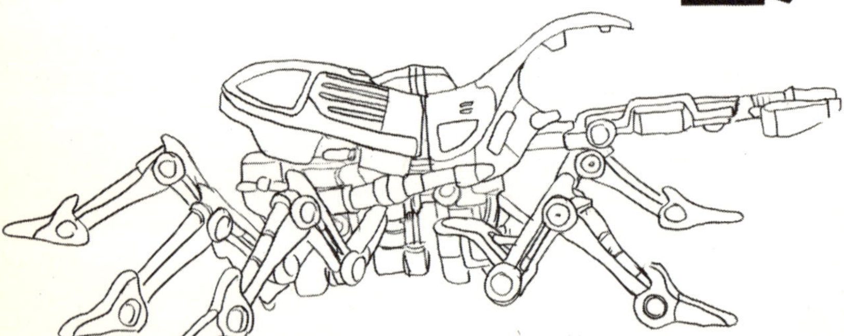

You have six legs here, so the middle pair of legs would be used to keep the center mass stable when the forelegs and hindlegs are in mid-stride.

The armor placement would correspond to the beetle's, protecting sensitive equipment -- in this case, the mech's engine.

The abdomen is segmented for movement.

You could make the beetle's horns into a sensor array and gun combination.

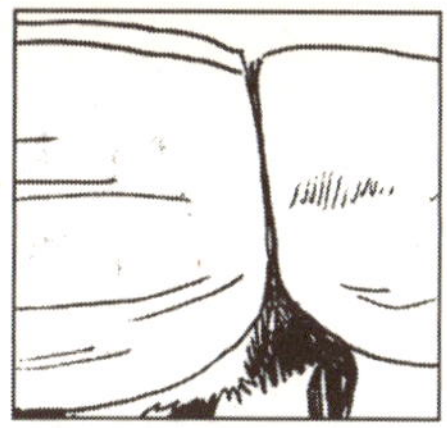

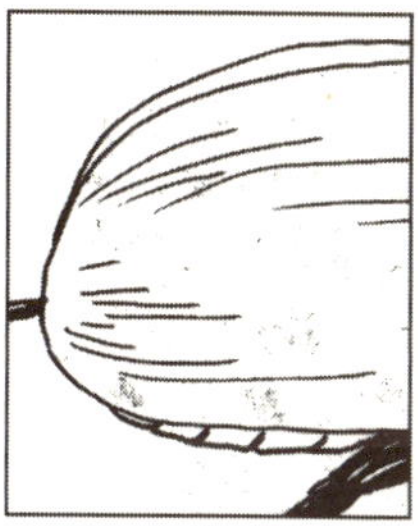

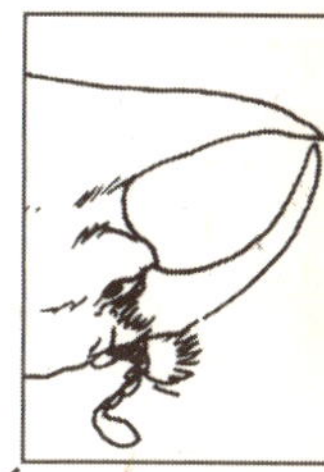

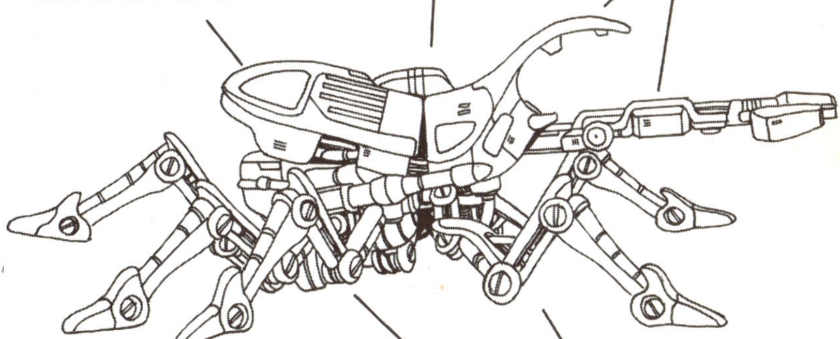

From an engineering standpoint, the legs are minimally jointed to provide essential movement.

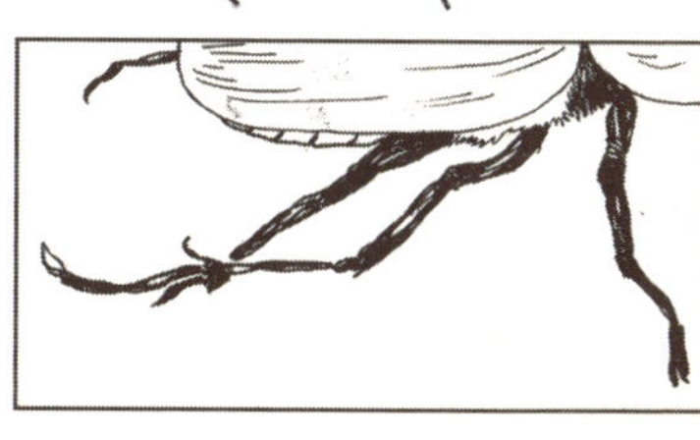

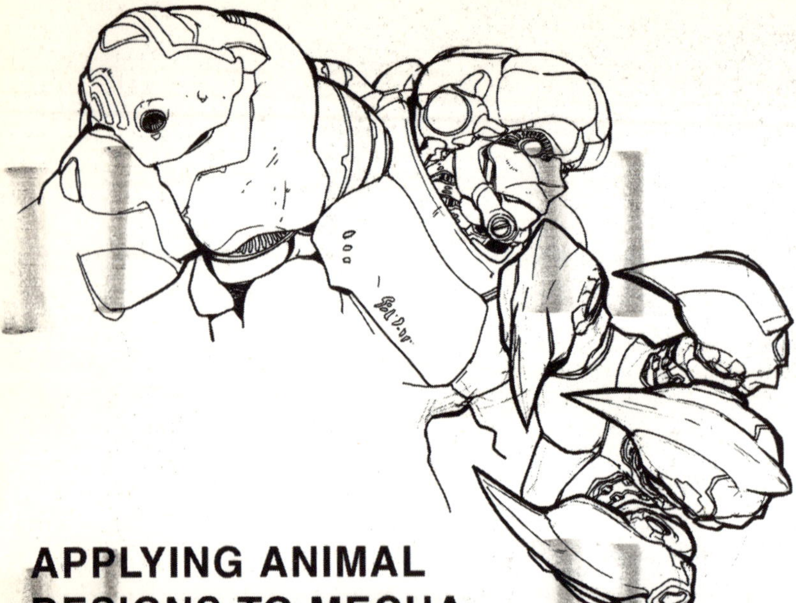

APPLYING ANIMAL DESIGNS TO MECHA

The mecha here looks like a regular bipedal mech, but has a hint of tiger qualities in its design, primarily in the claws and forehead.

Claw design

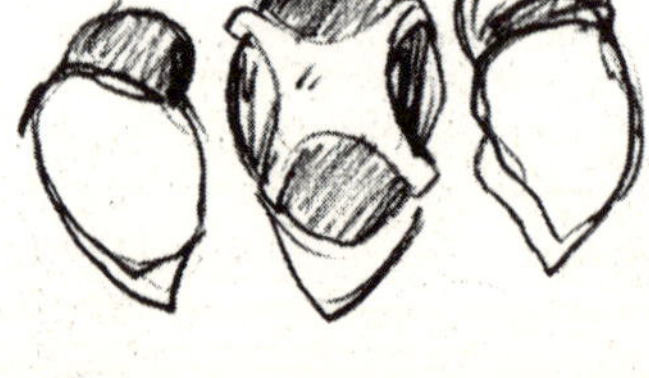

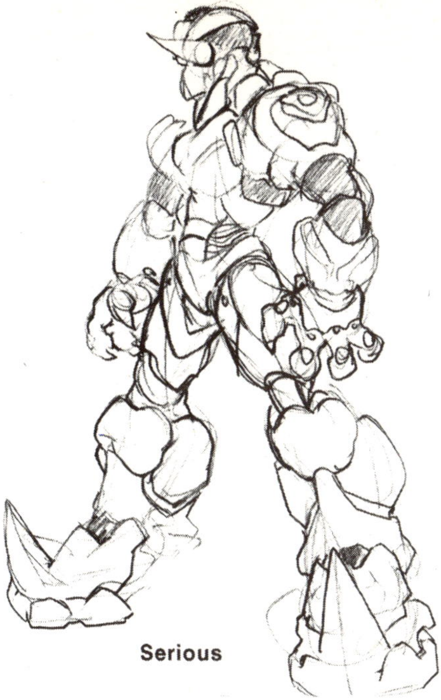

Serious

Merging designs isn't strictly for serious mecha. You can have fun with it.

The mecha here is meant to look cute and comical. It's an amalgamation of a cat and a monkey.

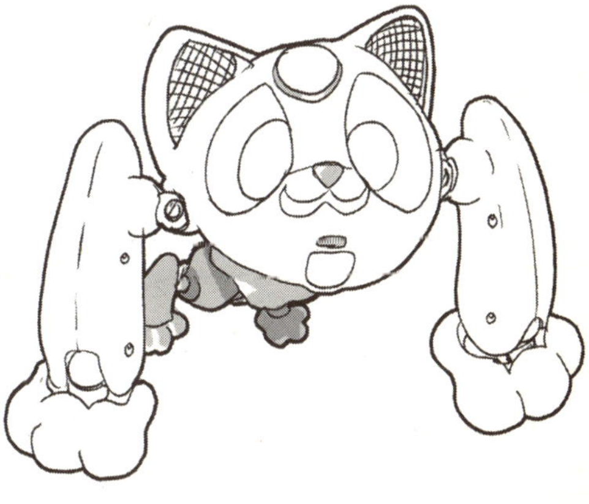

Comical

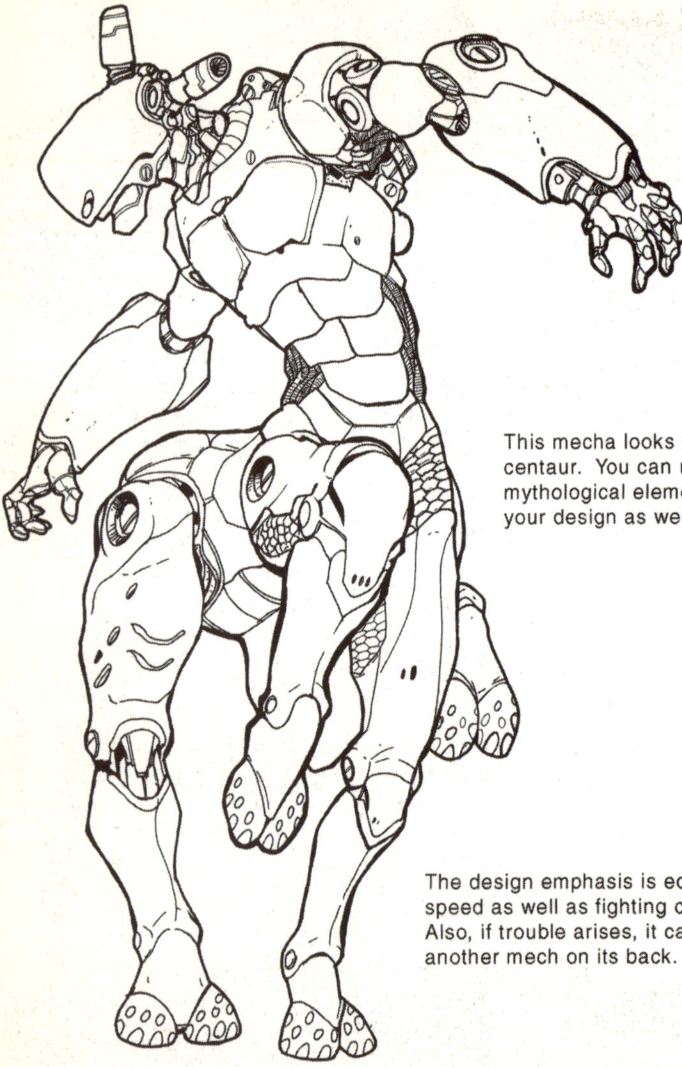

This mecha looks like a centaur. You can use mythological elements in your design as well.

The design emphasis is equally on speed as well as fighting capability. Also, if trouble arises, it can carry another mech on its back.

CHAPTER 08

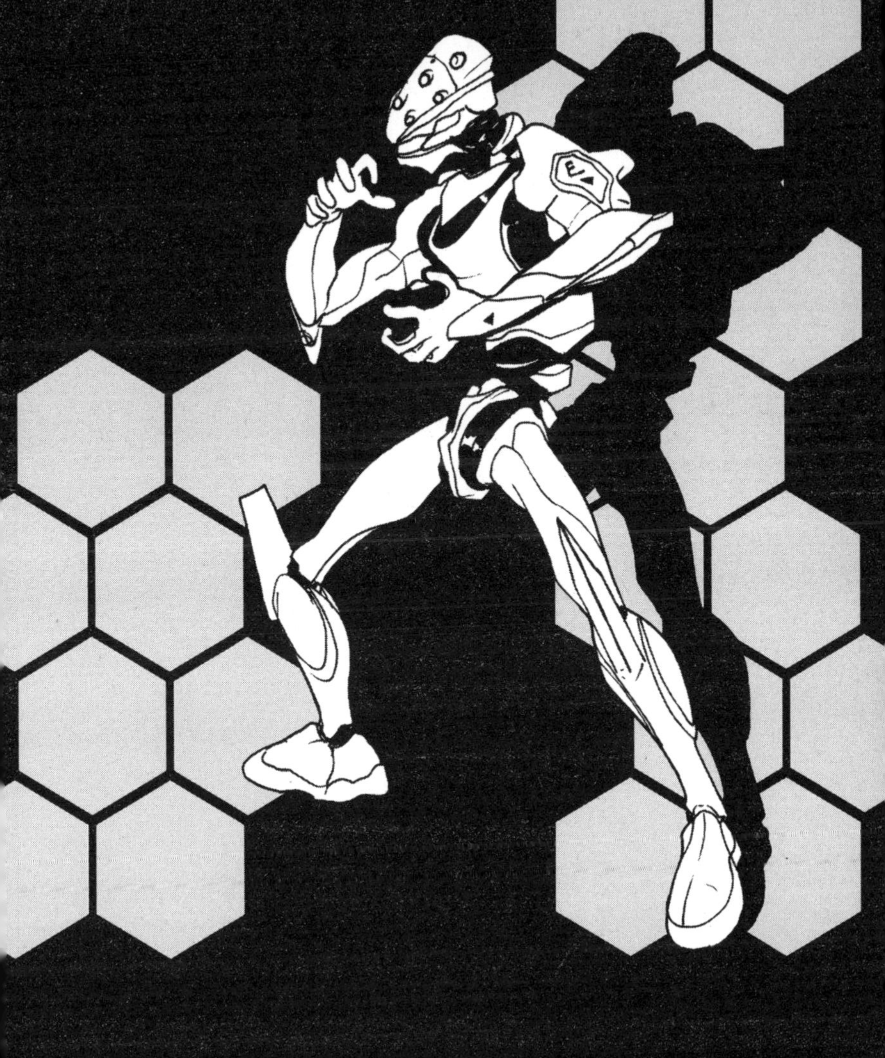

Organic Mecha

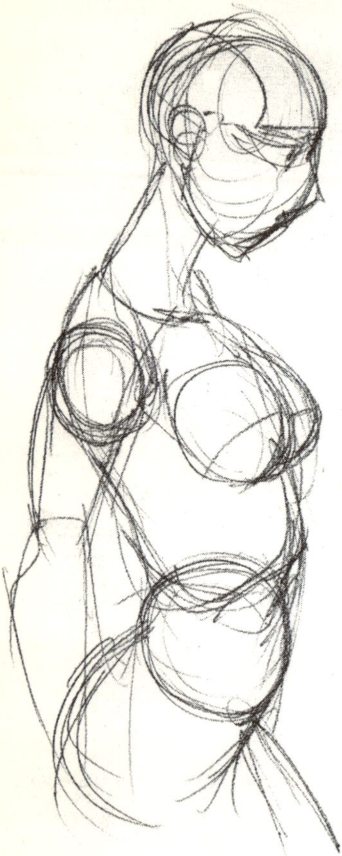

CYBORGS

A cyborg is a living organism surrounding an artificial interior, or a living thing with enhanced artificial parts.

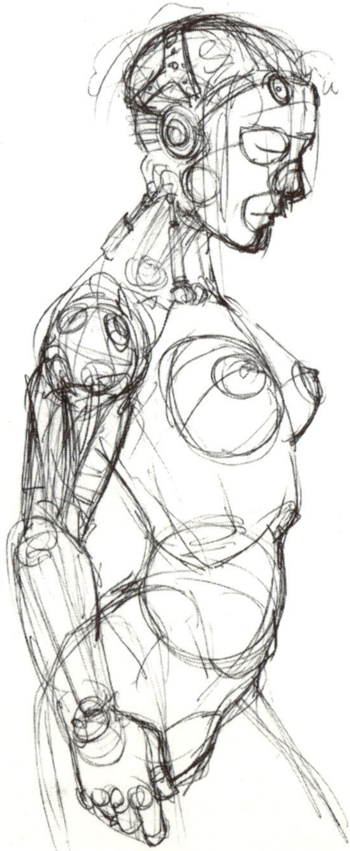

To draw cyborgs with a human exterior and a mechanical interior, start with a human form and map out the interior details.

Understanding anatomy will help you apply a mechanical basis to the cyborg's musculature.

DETAILS

Reference vehicle and computer parts when making the body. When you examine how a particular part operates and why it's needed, then you can apply that sensibility to your cyborg.

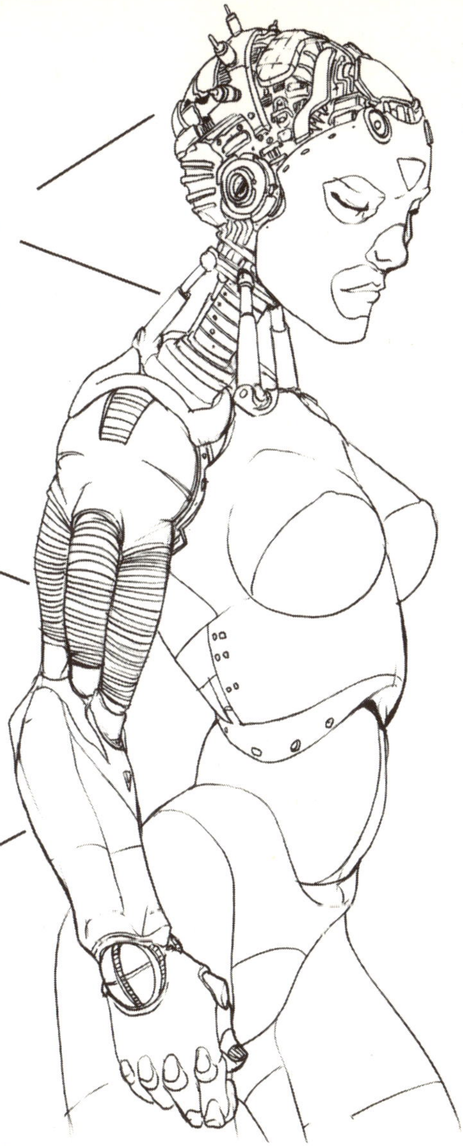

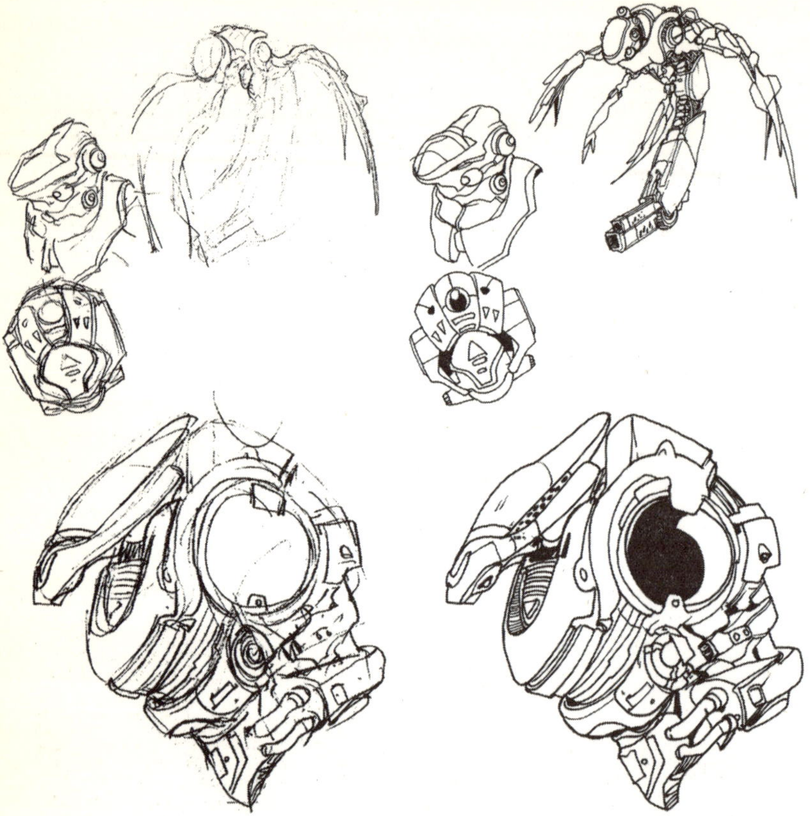

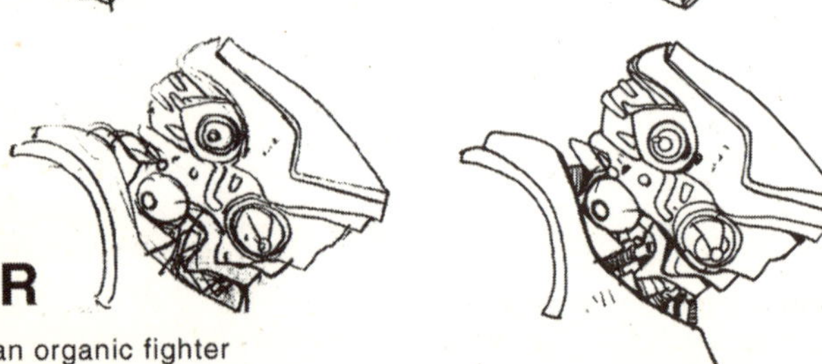

MECH FIGHTER

When creating an organic fighter mech, you can reference the structure of insects for its design.

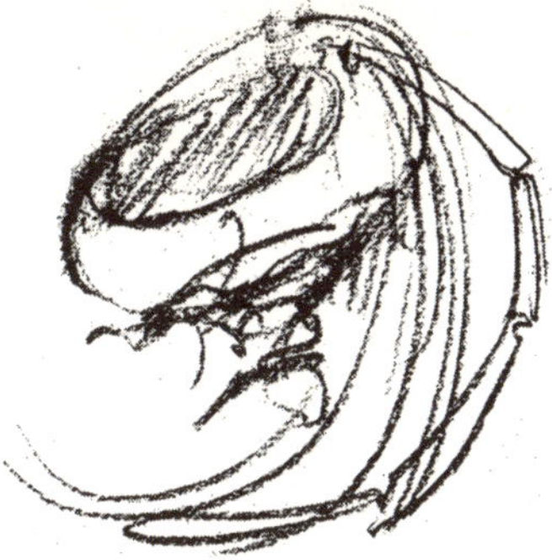

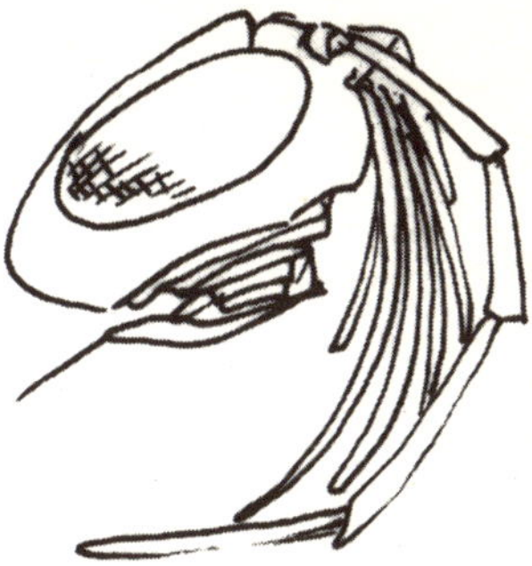

You'll want to figure out the personality of the mech. In this case, we're basing the head on an ant.

THE HEAD

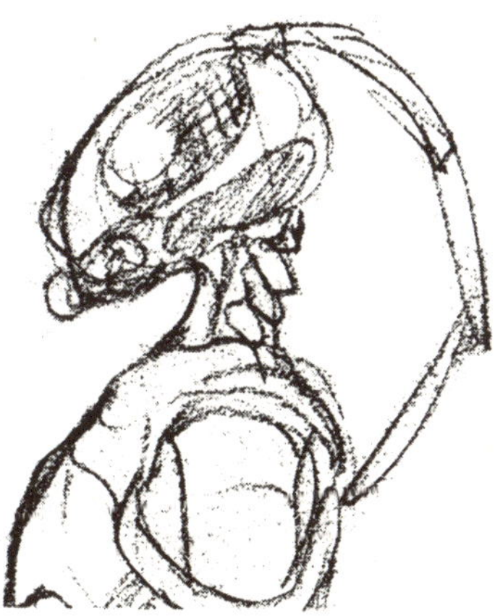

With a little more consideration, the head begins to take shape.

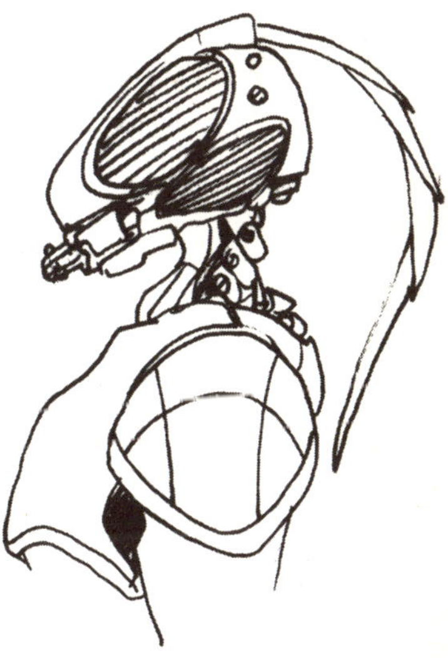

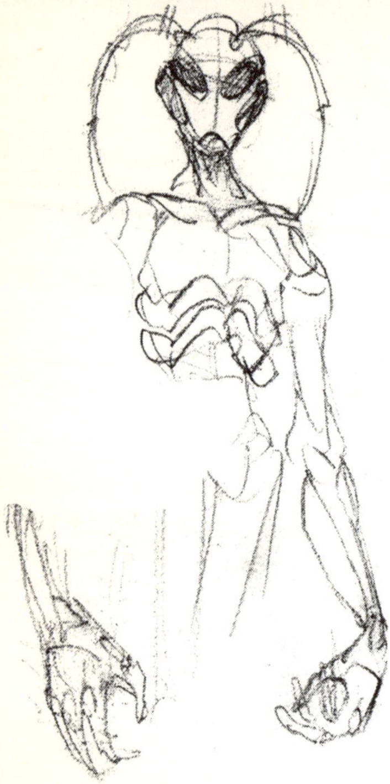

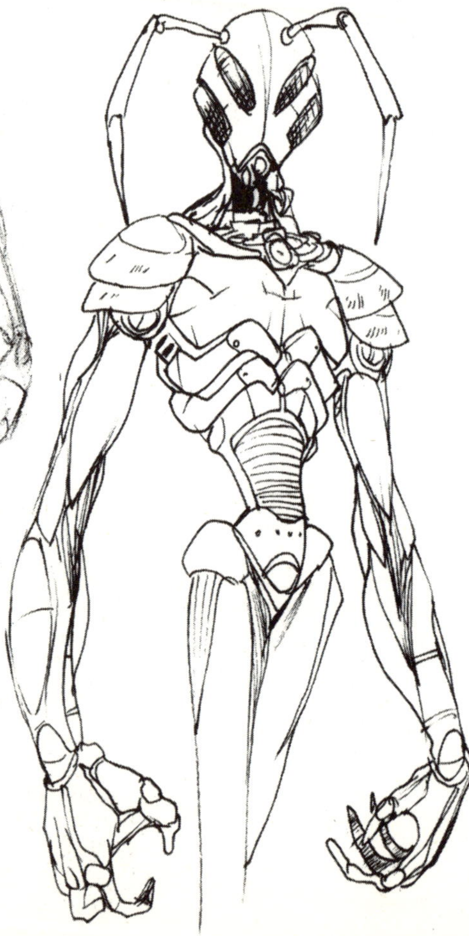

UPPER BODY

The upper body has a biomechanical look. In the early stages of design, the body is exceptionally thin.

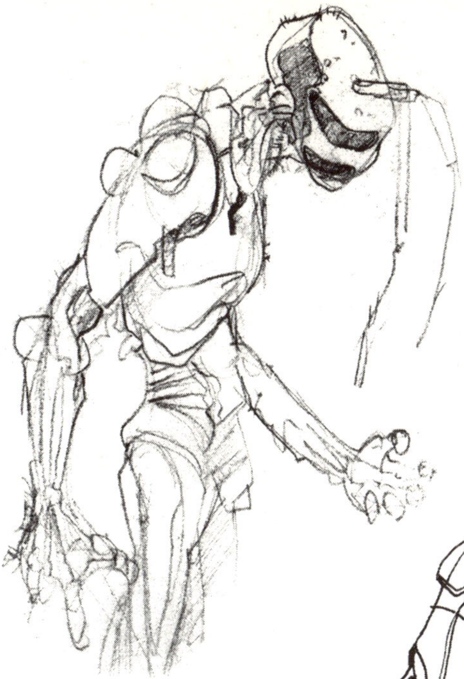

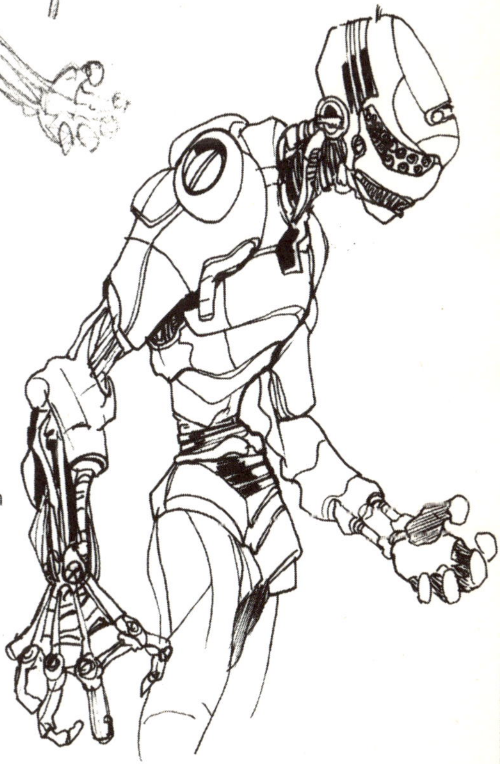

You can plan out two different designs: one with antennae and one without. Feel free to experiment with the mech's features to determine if certain aesthetics have a practical purpose.

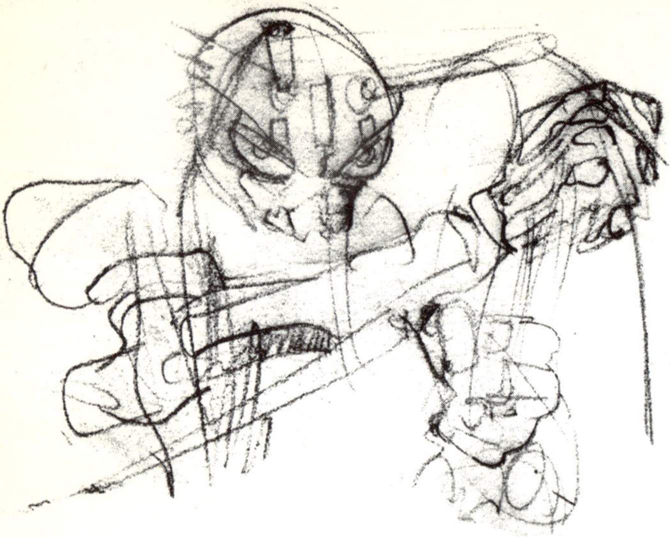

FOREARM

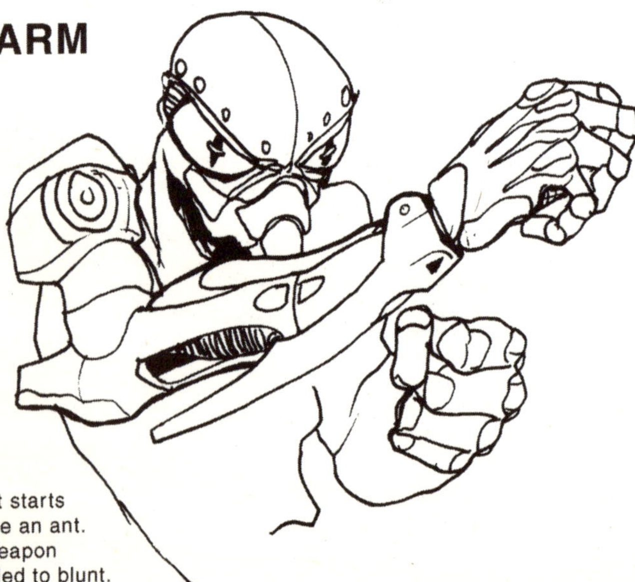

At this stage, it starts looking less like an ant. The forearm weapon goes from bladed to blunt.

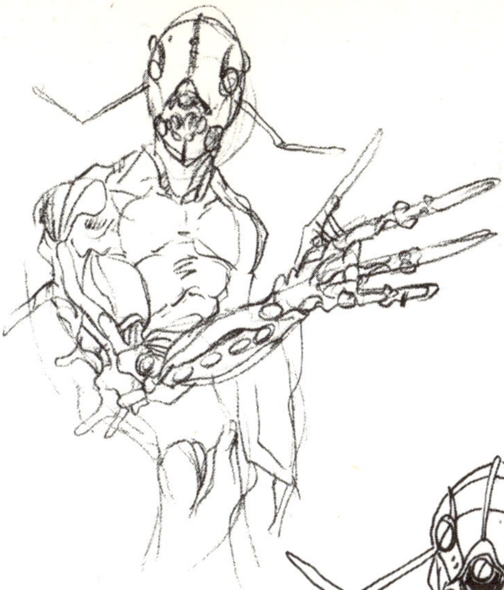

The body shape needs more clarity.

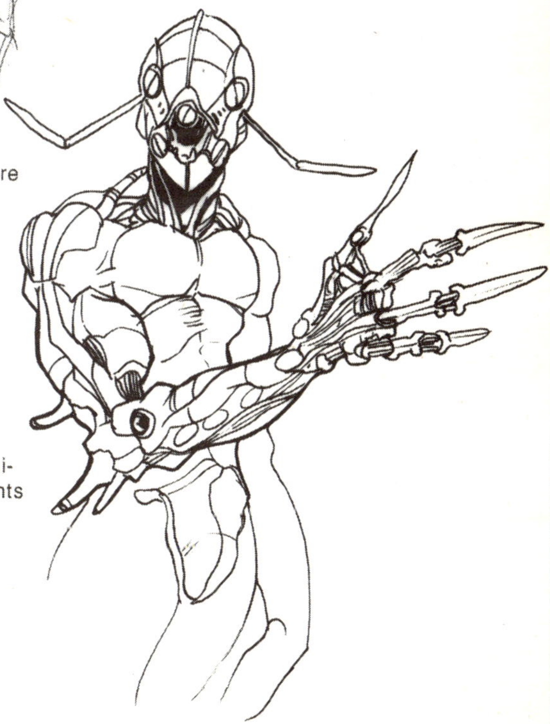

You want to put more definition in the arm. It represents the primary weapon of the fighter.

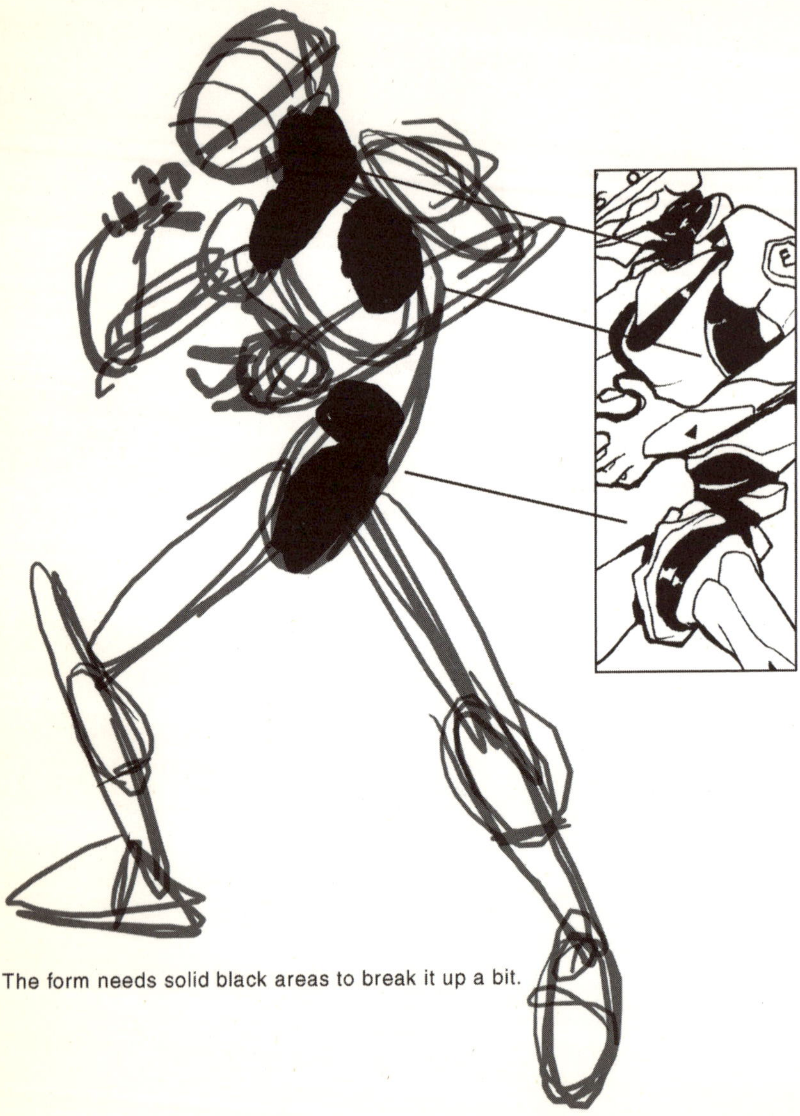

The form needs solid black areas to break it up a bit.

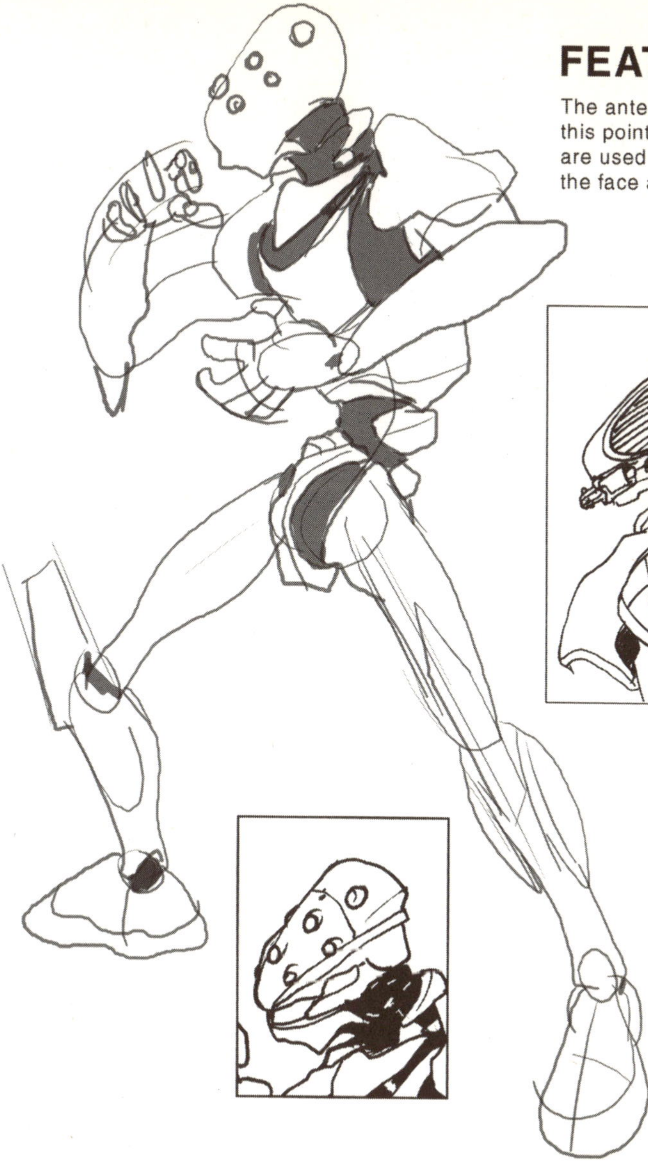

FEATURES

The antennae are gone at this point. Multiple eyes are used instead to give the face an insectoid look.

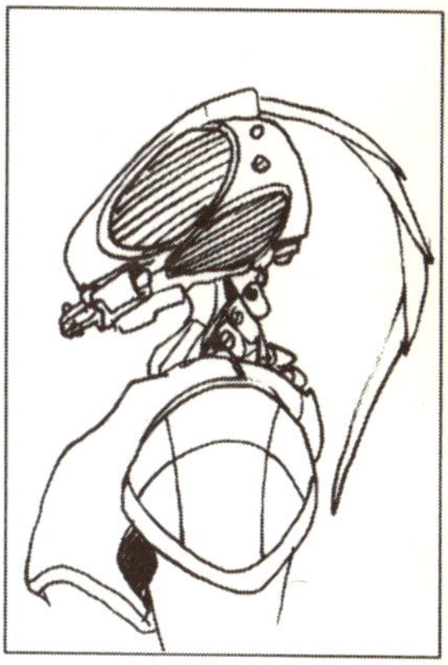

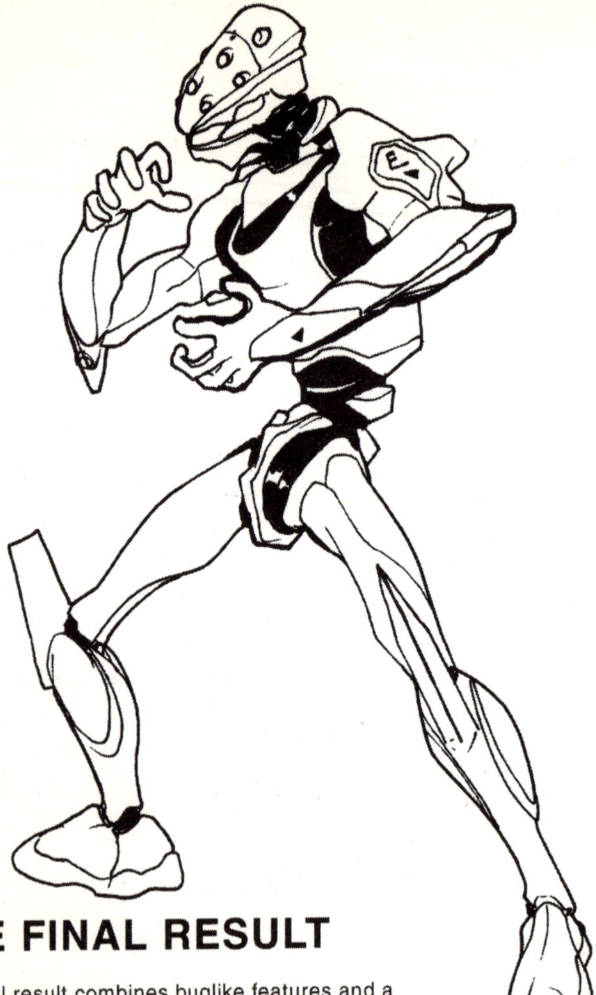

THE FINAL RESULT

Our final result combines buglike features and a streamlined, humanoid look.

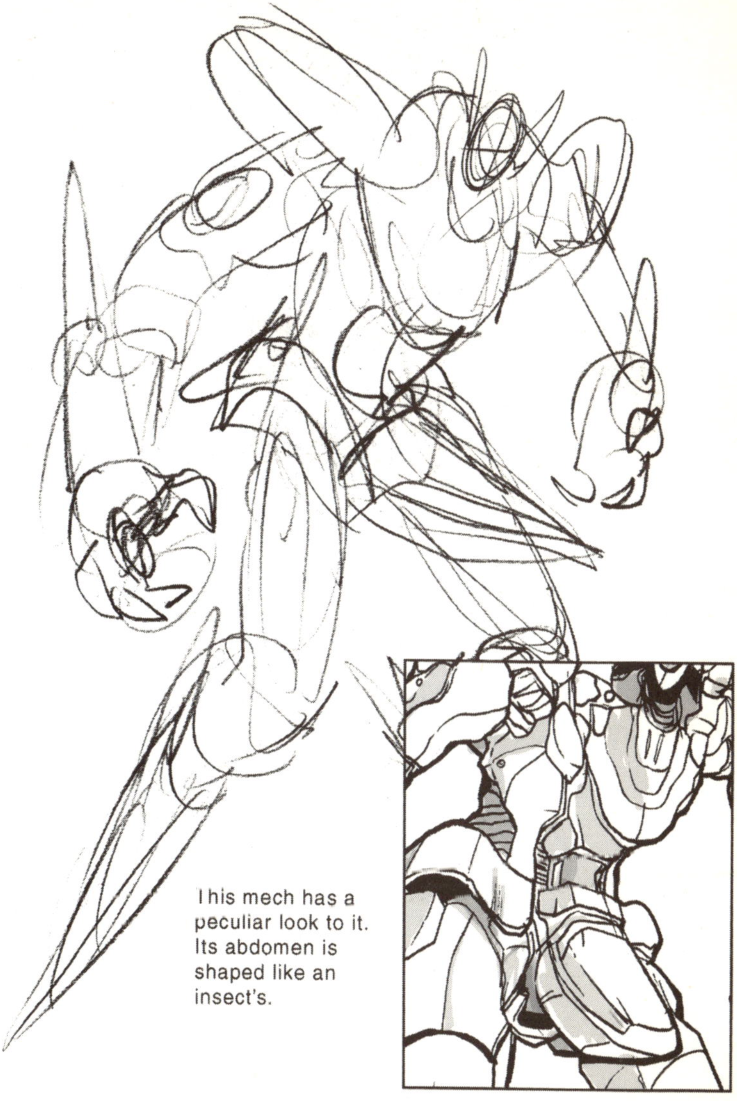

This mech has a peculiar look to it. Its abdomen is shaped like an insect's.

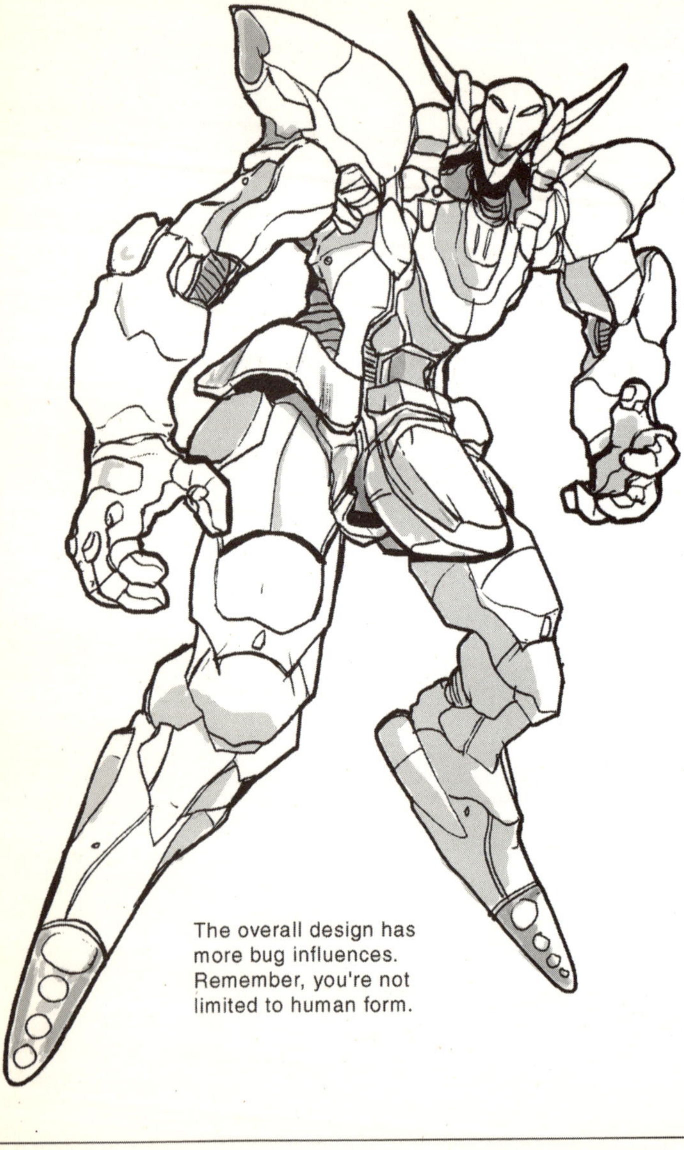

The overall design has more bug influences. Remember, you're not limited to human form.

With organic design, you want to keep the lines flowing. The fewer sharp angles you have, the better.

In this piece, the sharper edges are focused on the top of the head and to the back. These elements are positioned to flow into the rest of the curvy body.

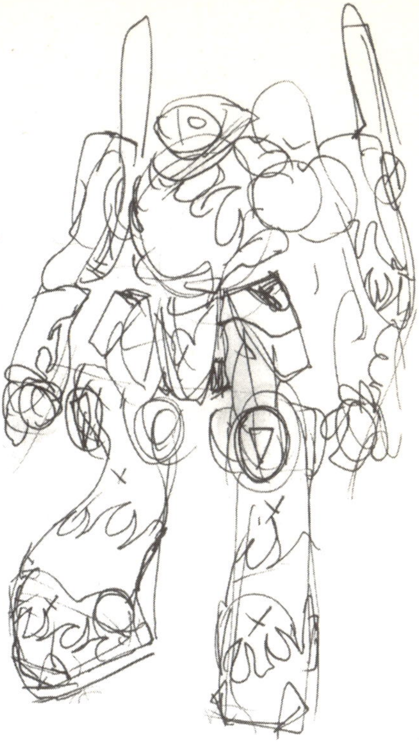

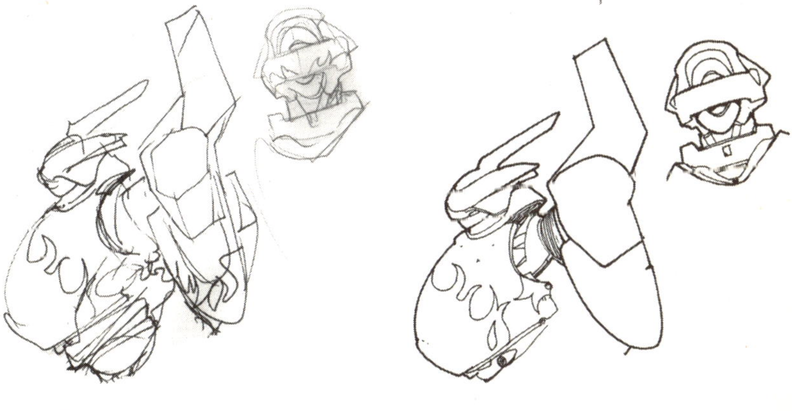

Key areas are filled in with black to help balance the figure. The flame pattern along its body shows that this mech is meant to be stylish.

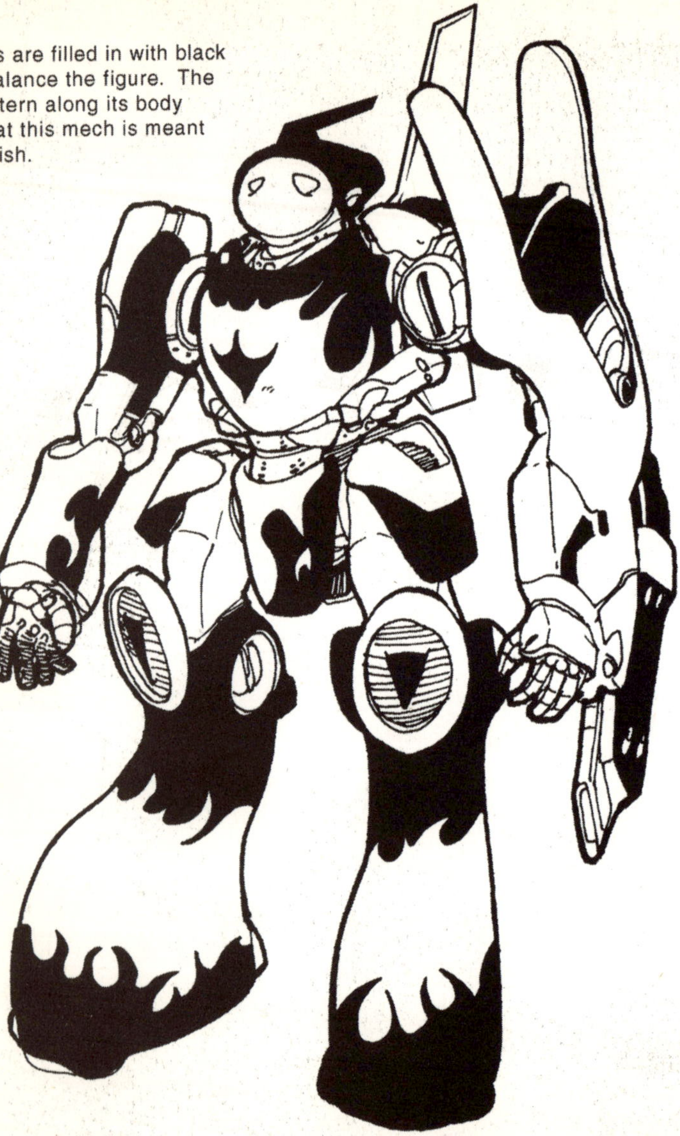

CHAPTER 09

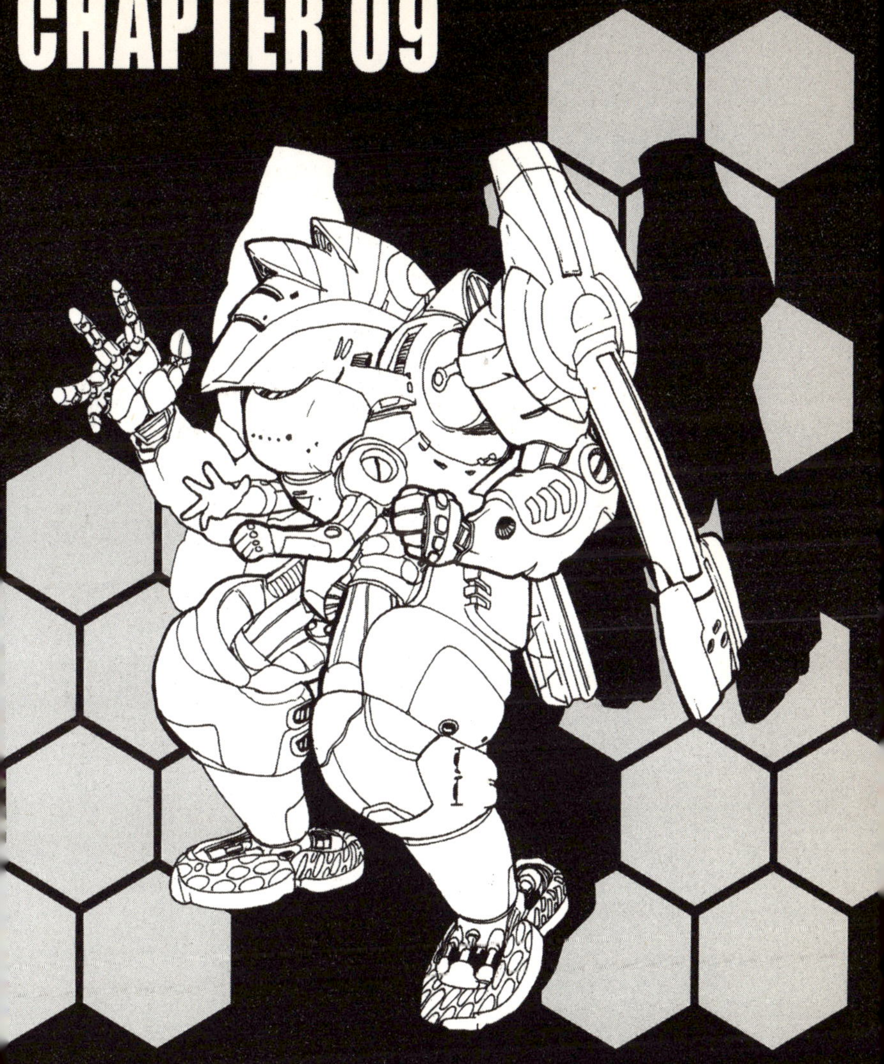

BATTLESUITS

MECHANICAL BODYSUIT SETUPS

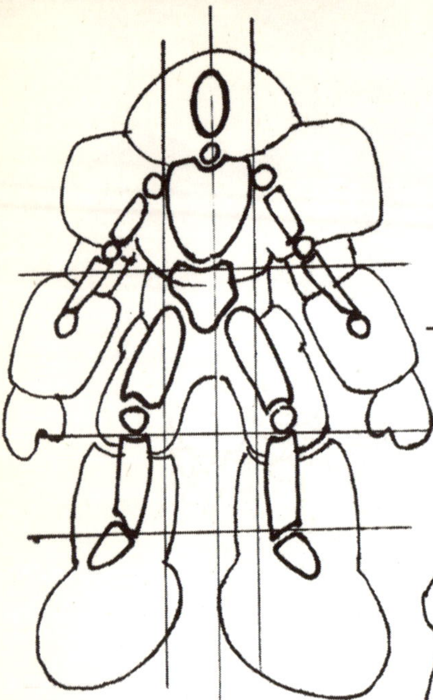

POWERSUIT

This is like a tank you'd wear, a mechanical suit of armor.

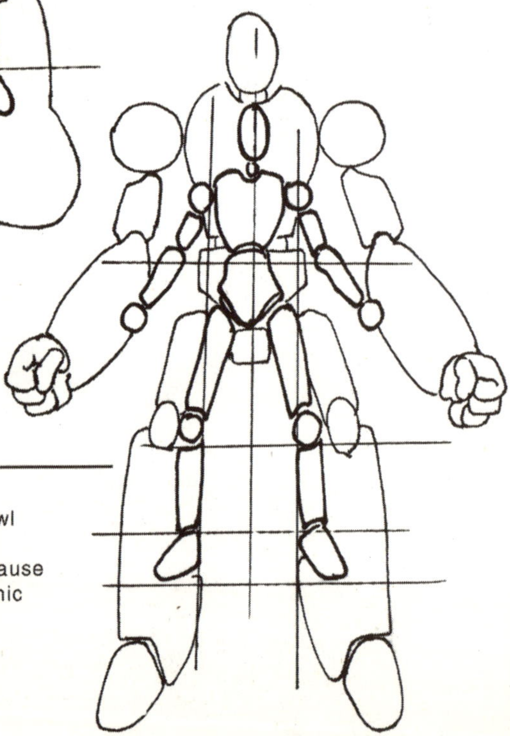

MASTER/SLAVE SUIT

This is a mech you can crawl inside of and operate. The physical controls you use cause the mechanical body to mimic your moves.

EXOSKELETONS

Exoskeletons are meant to enhance the physical attributes of the user. Depending on their design, they can be either power suits or master/slave suits, like this one.

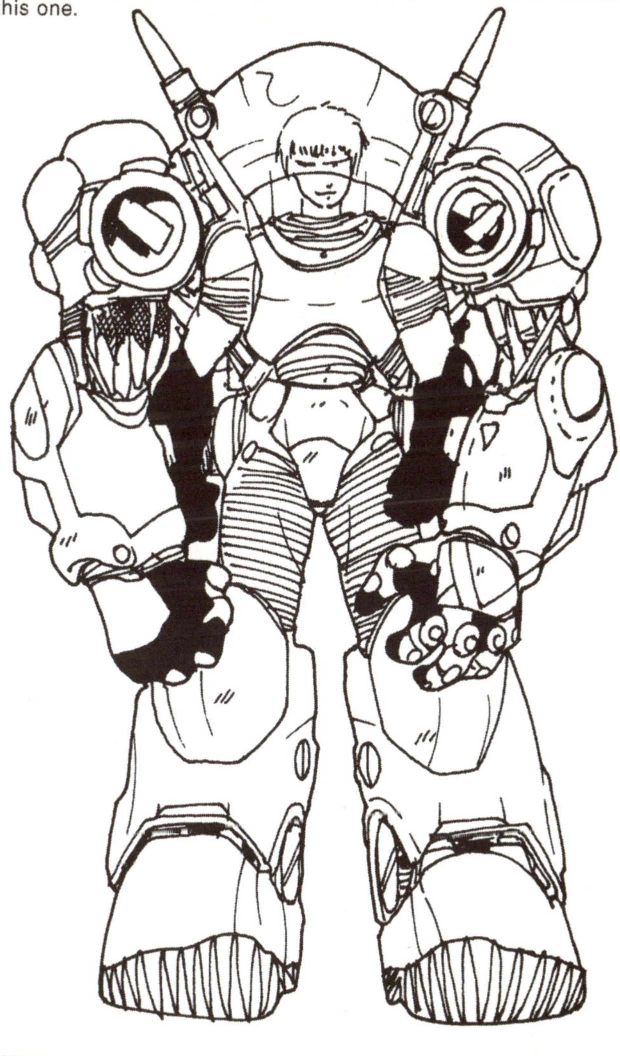

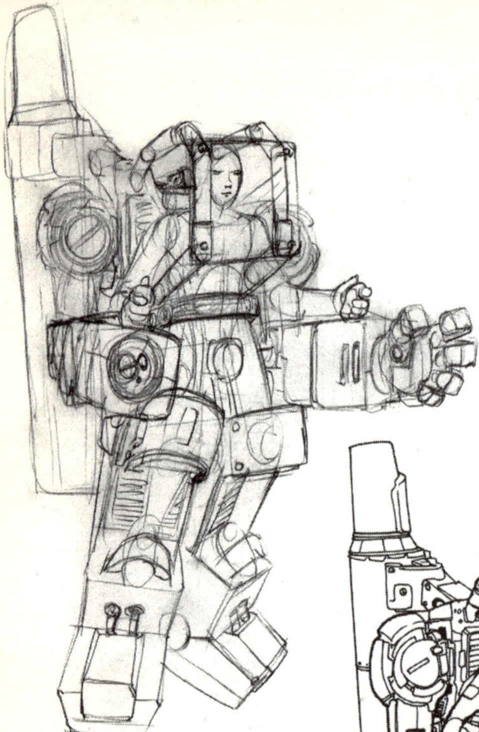

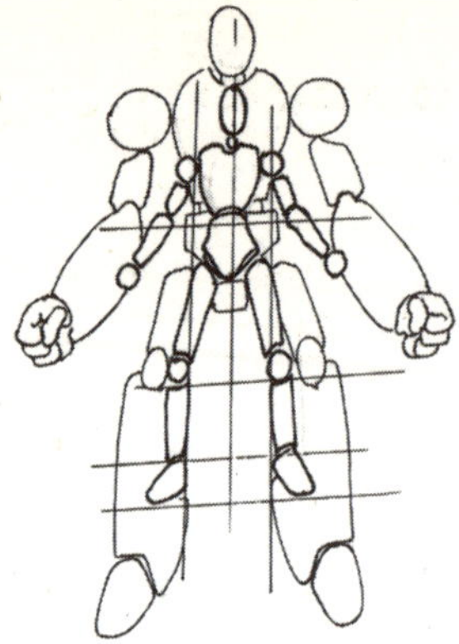

This is another master/slave suit example. The open compartment is made for easy access.

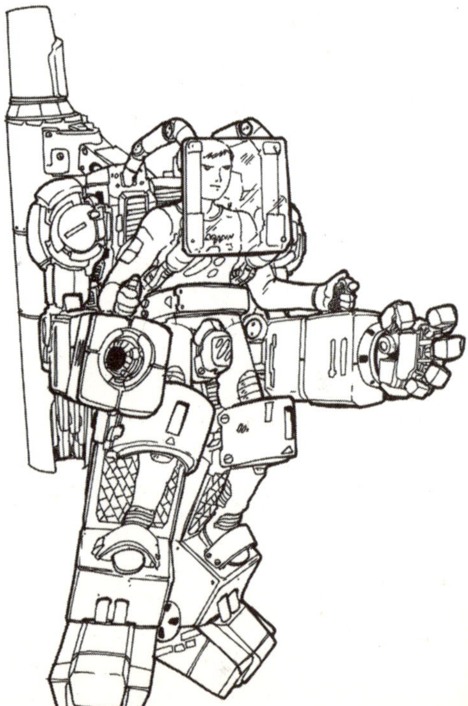

PRESENTATION

There's more detail in the foreground.

Put less detail in the background of the figure.

Minimal detail on the back arm makes the human arm stand out.

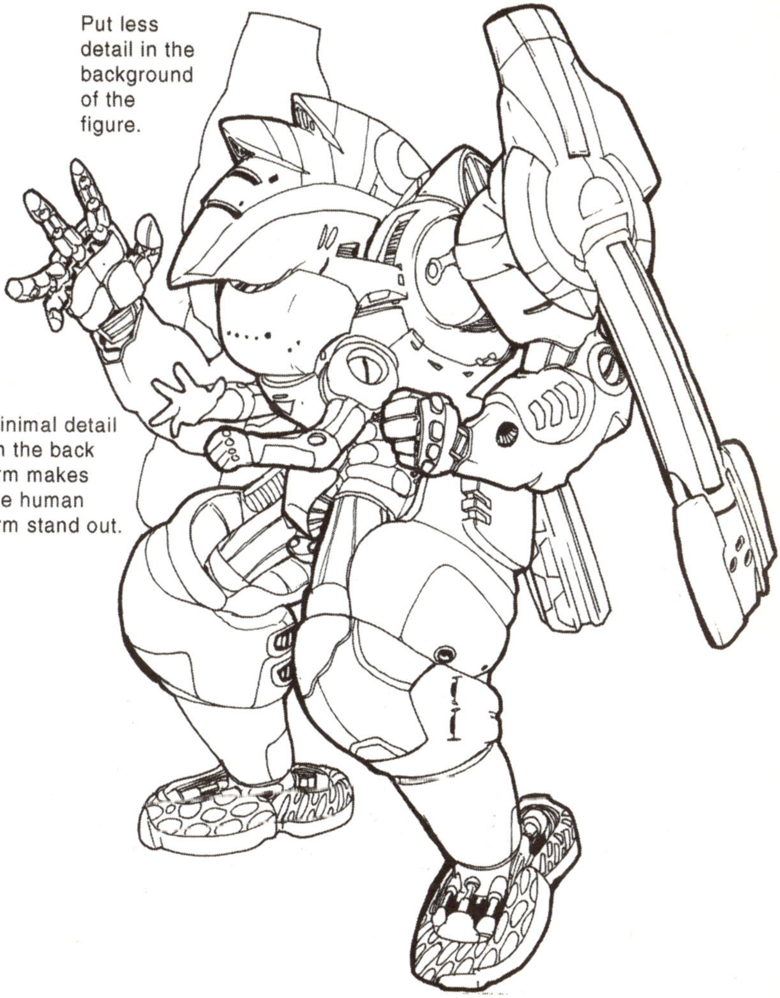

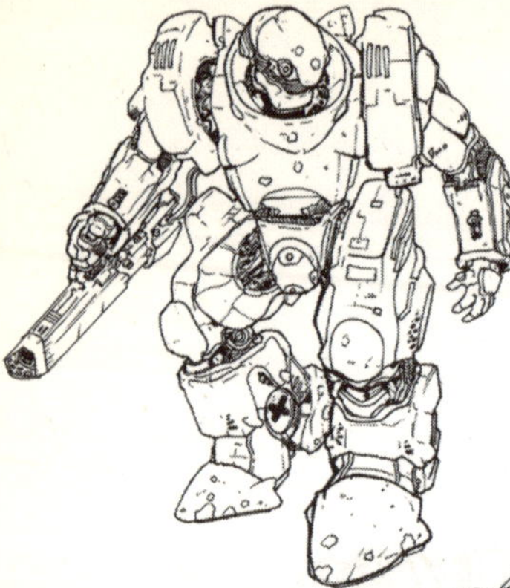

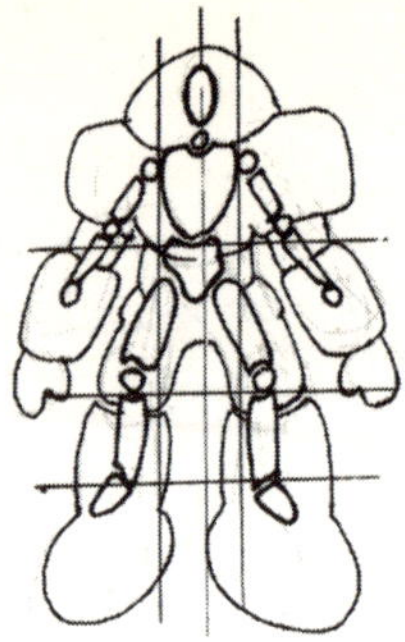

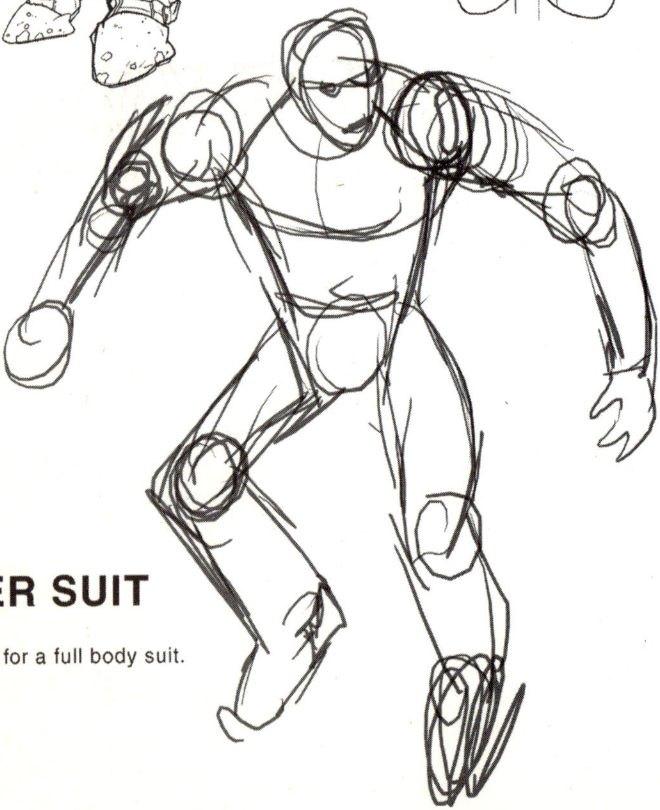

POWER SUIT

Preparation for a full body suit.

The person is completely covered in the robotic shell. It has all the weapons and life support systems he'll need.

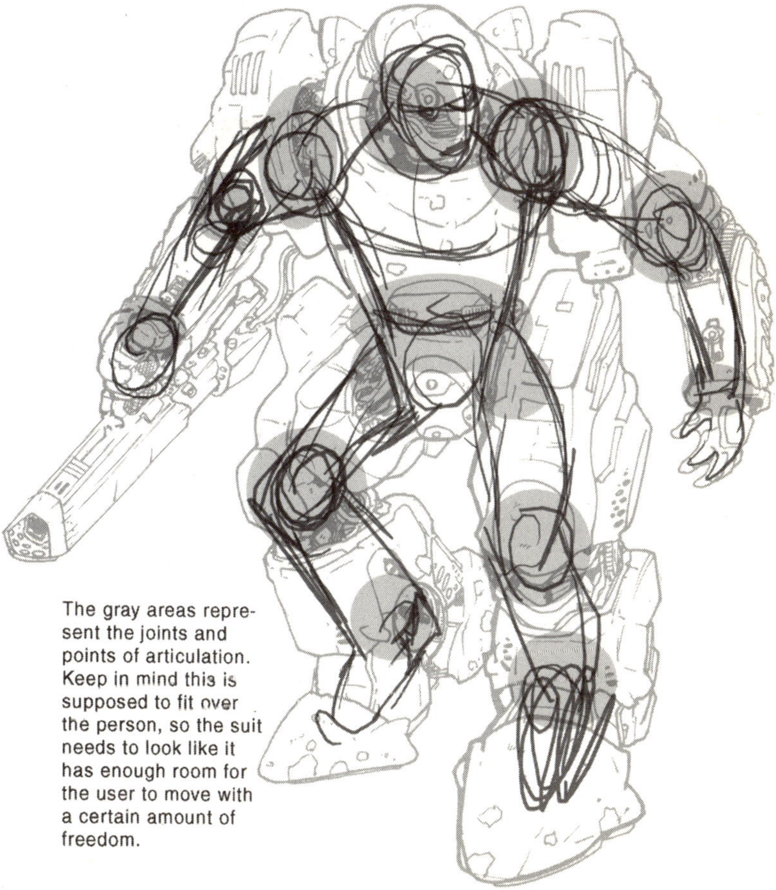

The gray areas represent the joints and points of articulation. Keep in mind this is supposed to fit over the person, so the suit needs to look like it has enough room for the user to move with a certain amount of freedom.

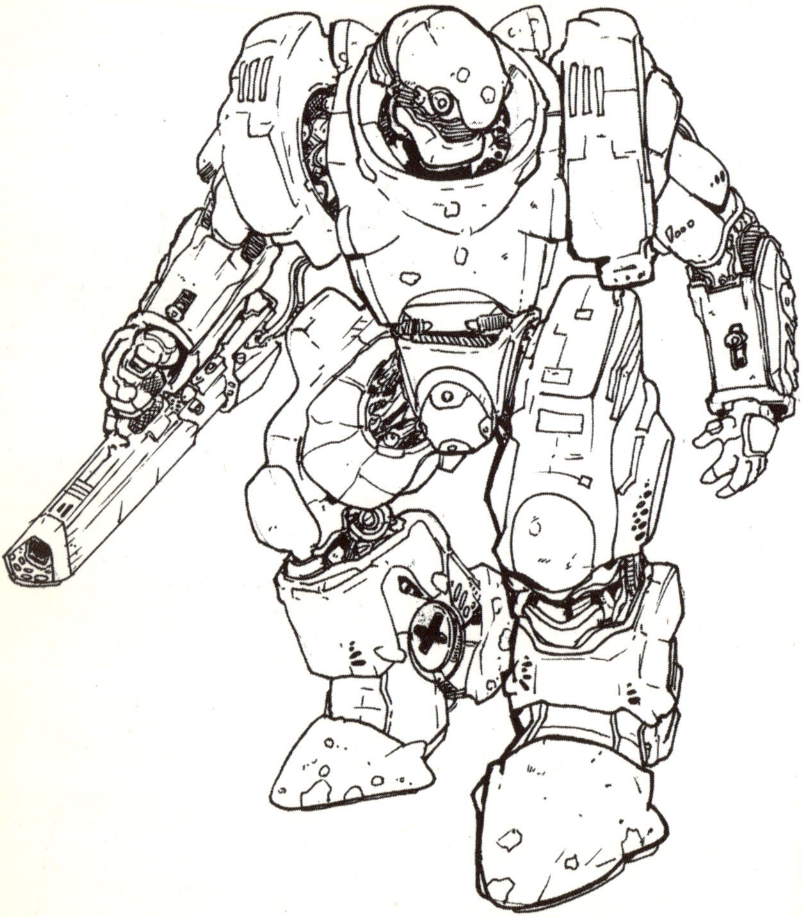

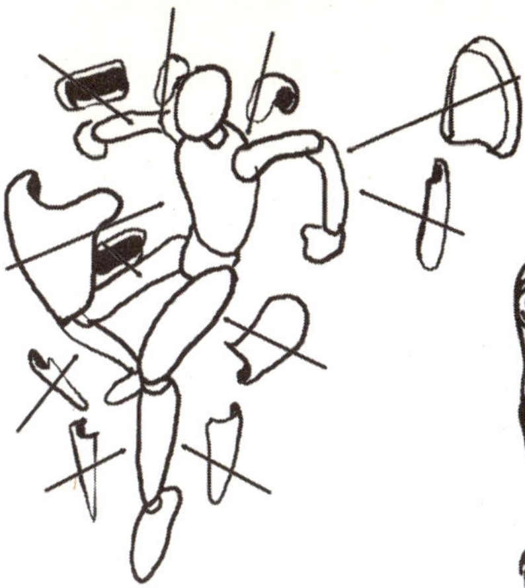

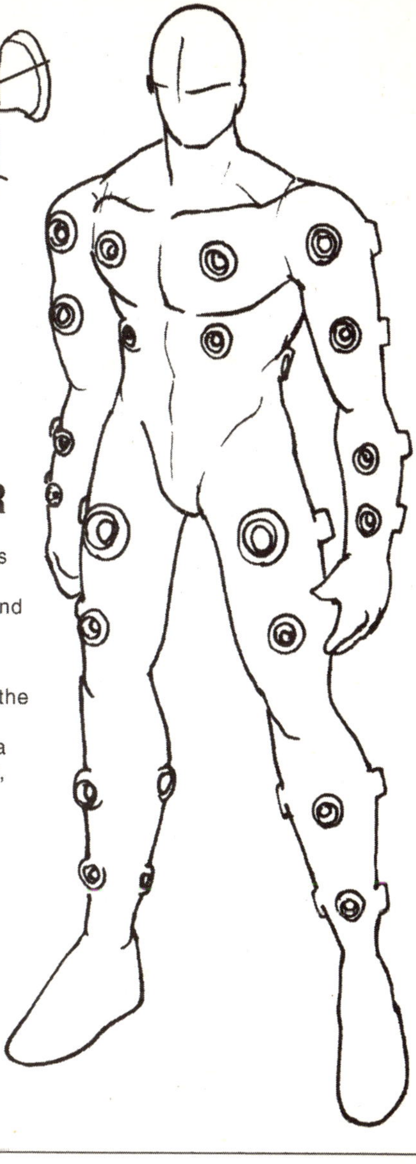

HYBRID ARMOR

This type of armor uses properties of both master/slave and power suits. The armor is wrapped around the figure.

To aide in the wrapping process, the bodysuit has armor connection ports all over it. You can design a myriad of suits and even vehicles, depending on the position of the body.

POSITIONING

Here are four simple forms for mechanical attachments. The body is in the positions it needs to be to accept the armor.

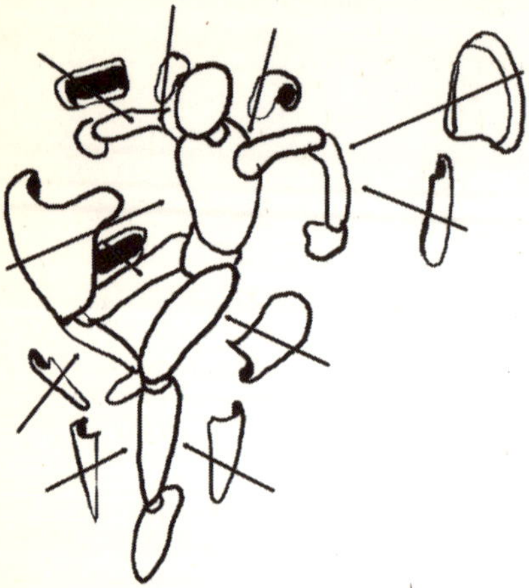

FLIGHT

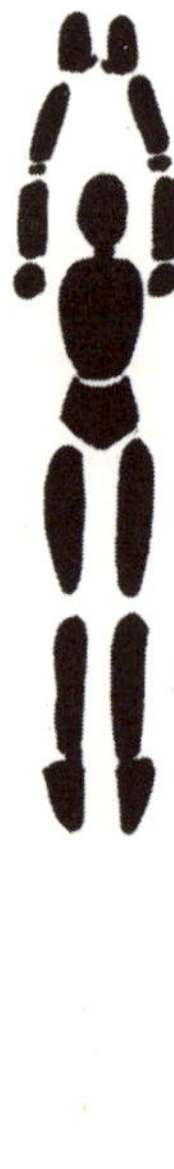

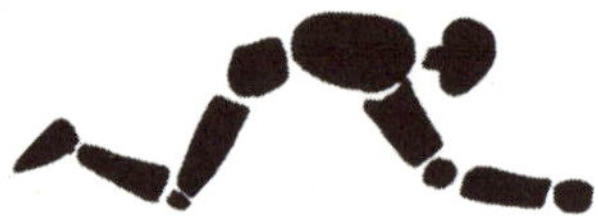

MOTORCYCLE

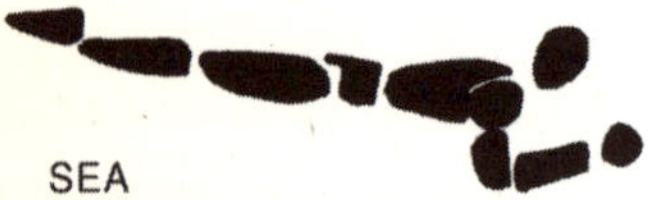

SEA

GUNNER

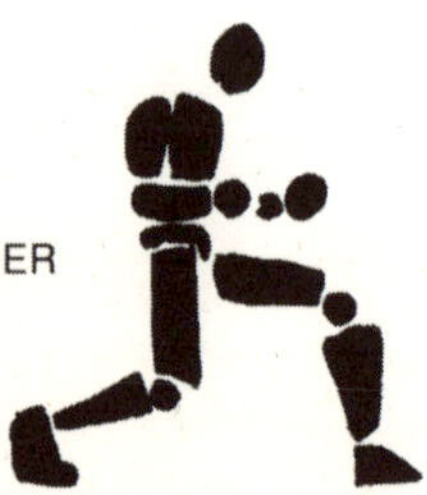

MOTORCYCLE

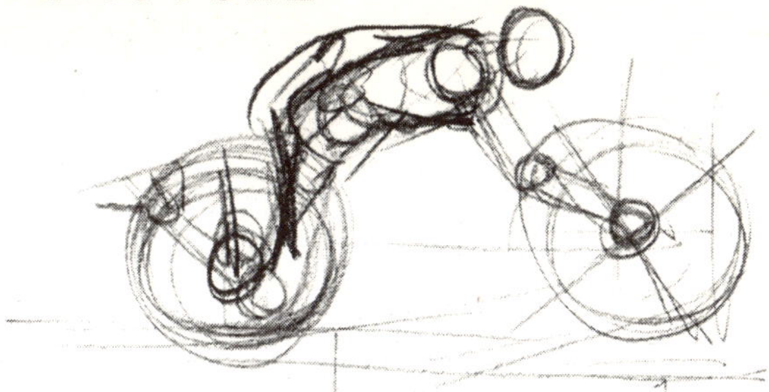

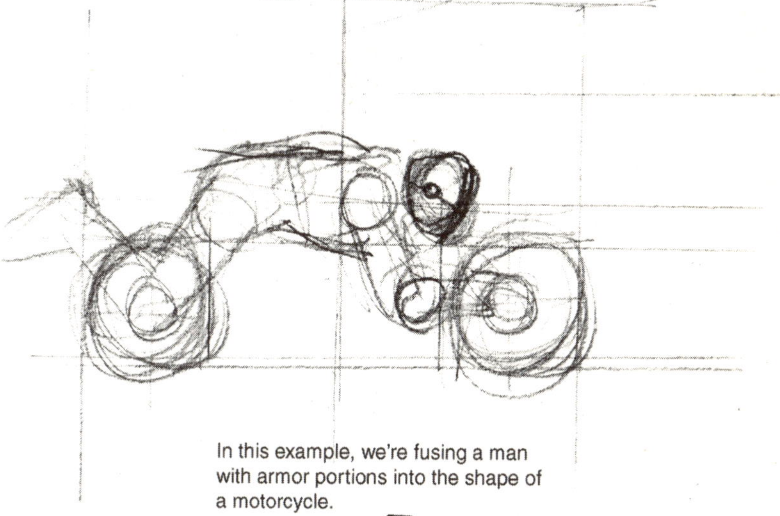

In this example, we're fusing a man with armor portions into the shape of a motorcycle.

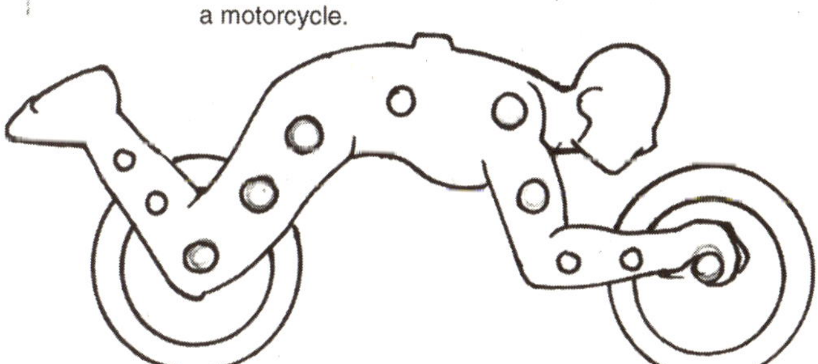

The armor and body suit are meant to give the impression of a motorcycle. The look doesn't need to be exact. When you're designing the motorcycle, consider how comfortable a human pilot would be inside of it.

Form

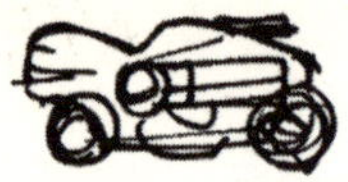

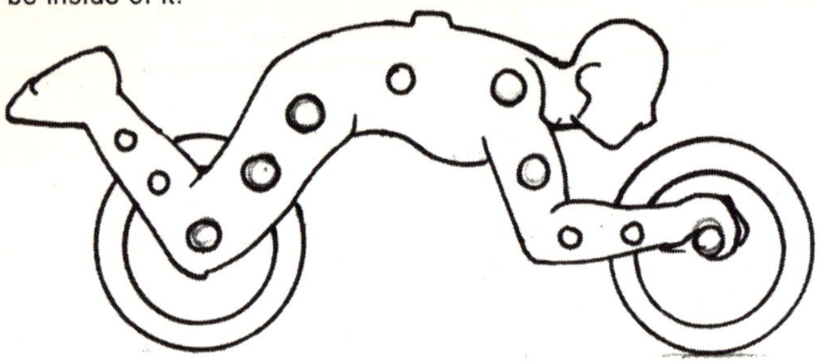

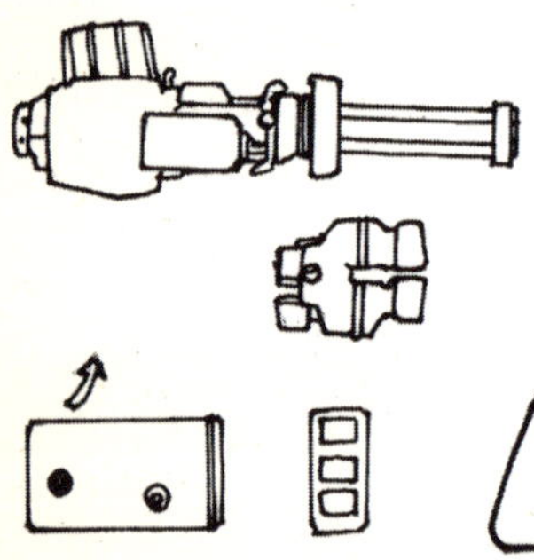

Here's a cache of parts and weapons that'll be applied to the figure to make the motorcycle.

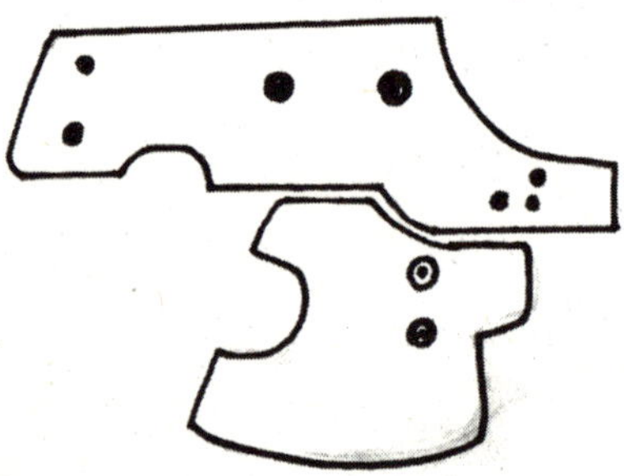

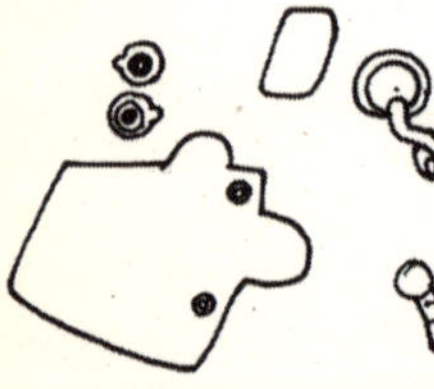

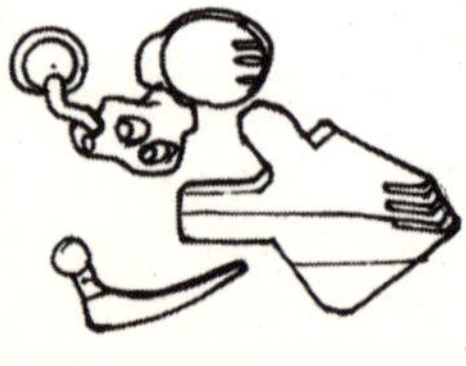

LAYERING

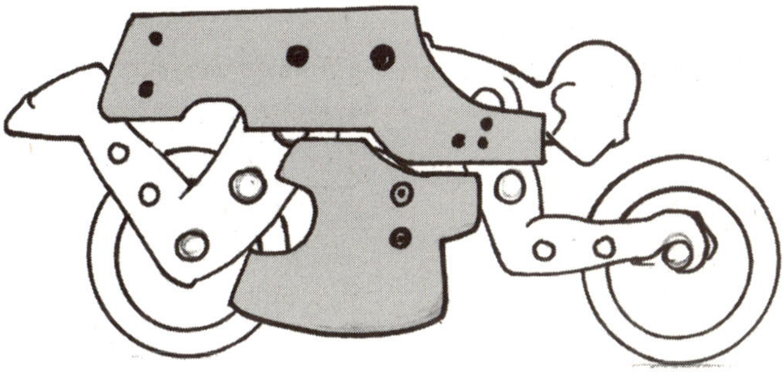

The armor should be pieced together in stages. When you're layering the body suit, the fundamental sections responsible for vehicle functions should be connected first. Atop that comes the equipment responsible for weapons control, and then the armor.

Try not to apply too much material onto the body. If you need to, decrease the number of portions or armor to fewer stages. You don't want the figure to look like it would collapse under its own weight.

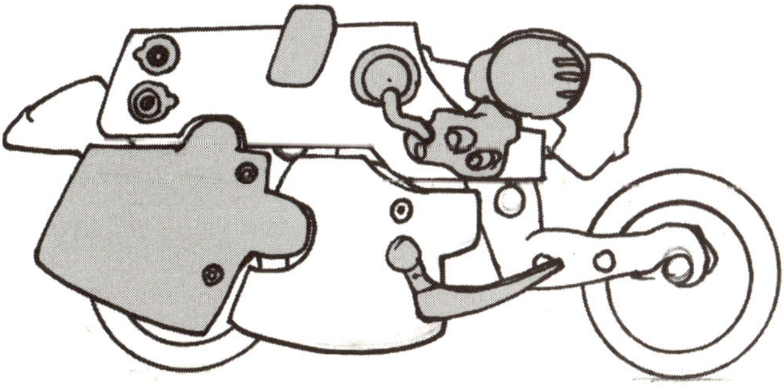

And finally, you place the weapons. The illustration here is a previsualization of the assembled armored vehicle.

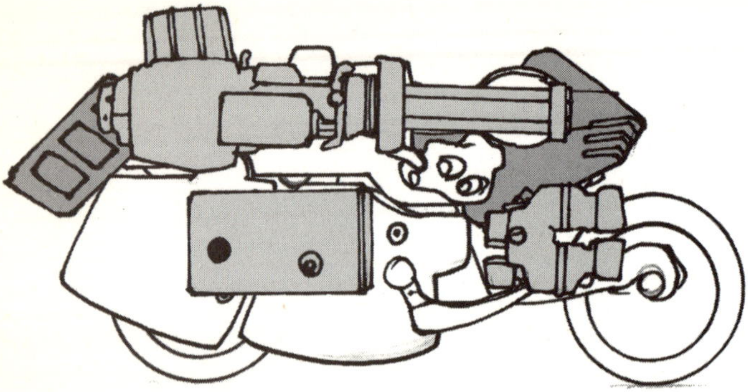

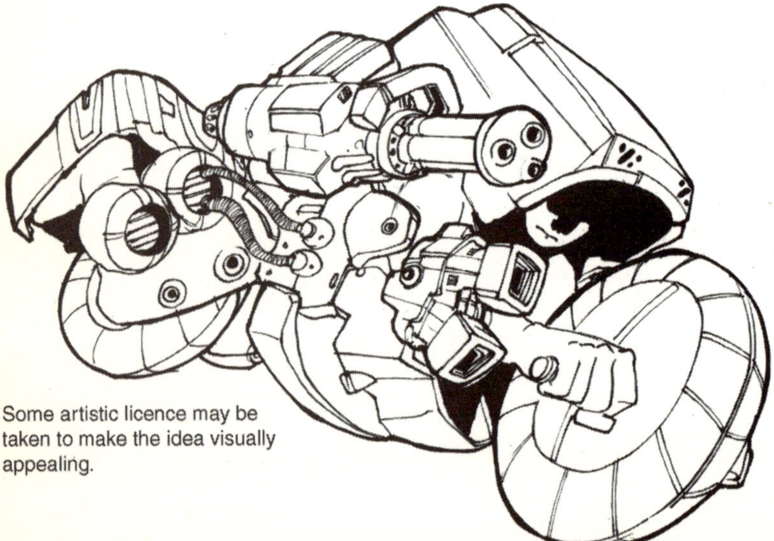

Some artistic licence may be taken to make the idea visually appealing.

PILOT SUITS

The pilot suit or flight suit gives a pilot protection and life support in combat. Just like with a mecha suit, you want functionality and comfort to work together. For believability, reference biker suits and take note of helmet design and padded areas.

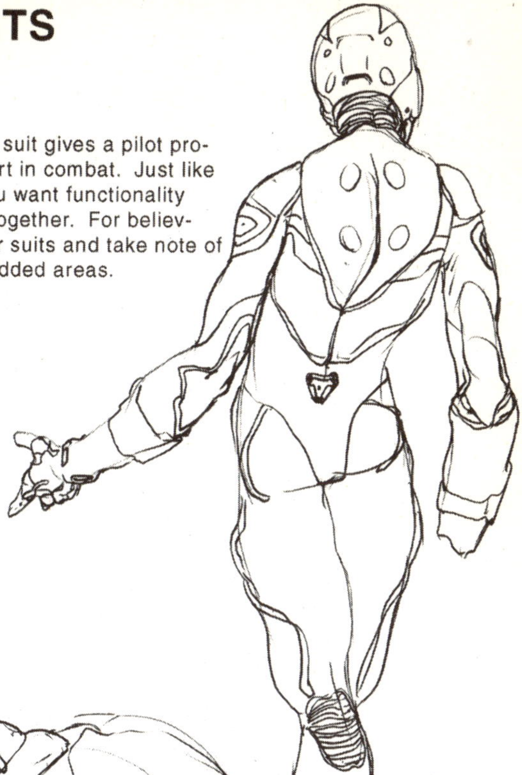

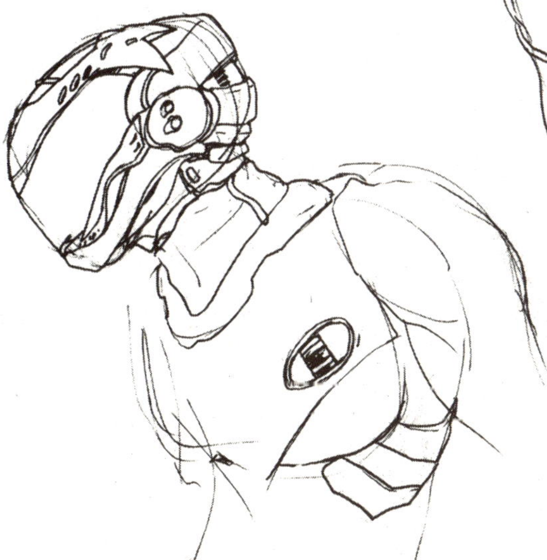

OVERALL STRUCTURE

When you create a pilot suit, you want to begin by building atop the body so that the suit looks like it will fit the pilot comfortably.

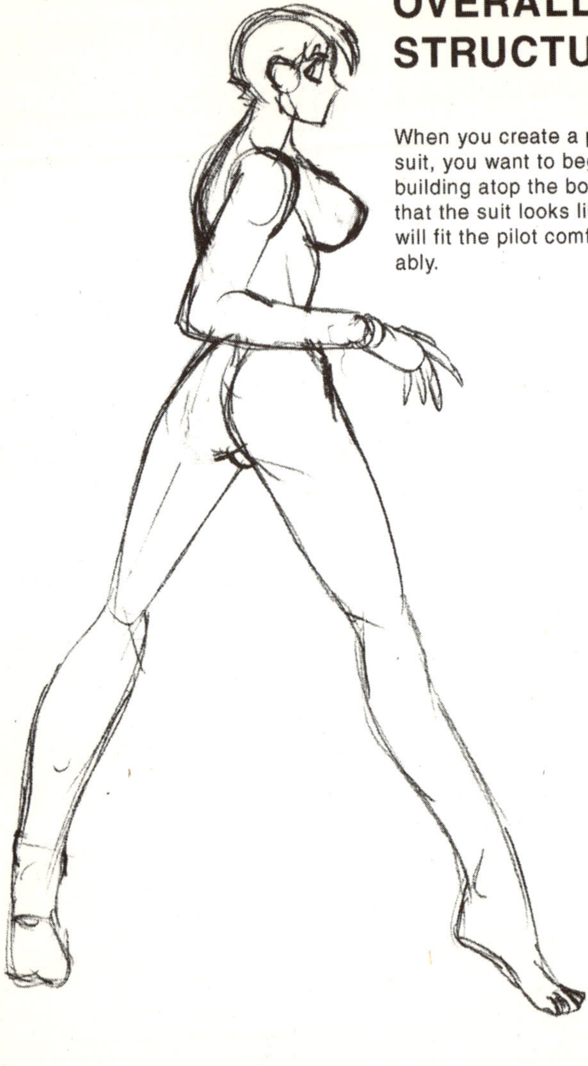

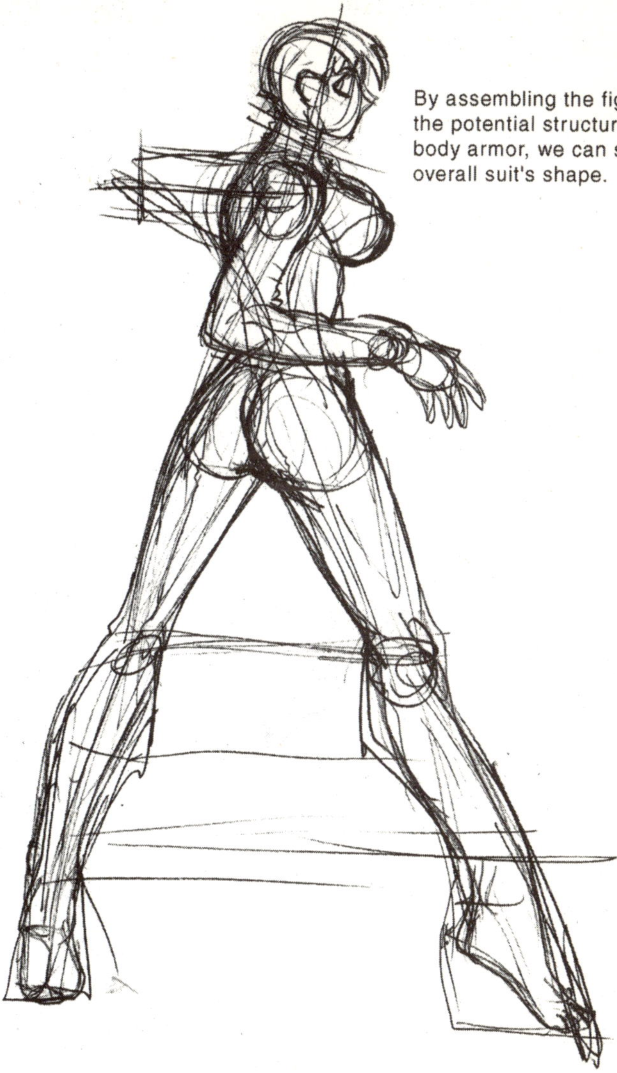

By assembling the figure with the potential structure of the body armor, we can see the overall suit's shape.

BODY SUIT

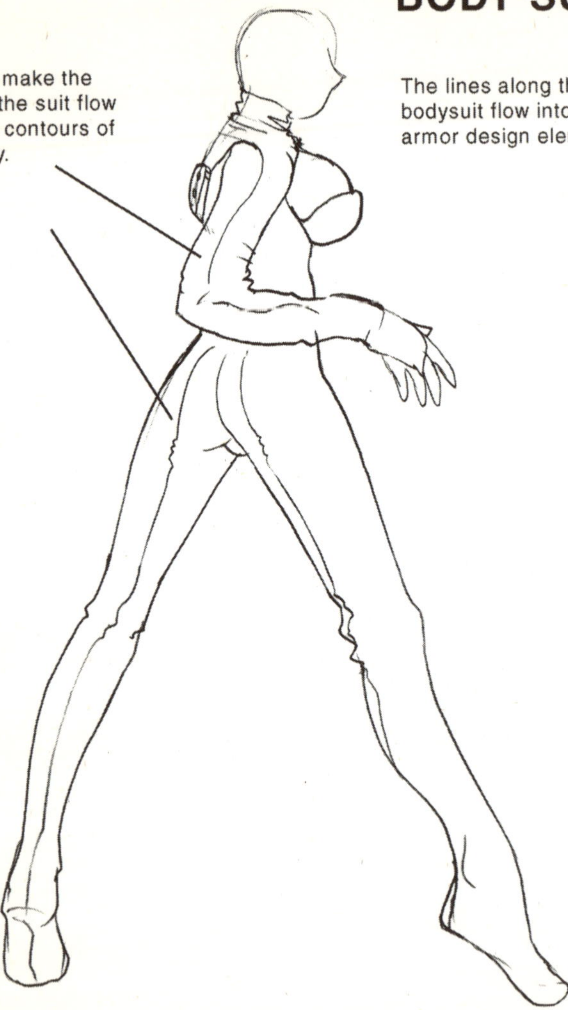

Try to make the lines in the suit flow with the contours of the body.

The lines along the bodysuit flow into the armor design elements.

ARMOR

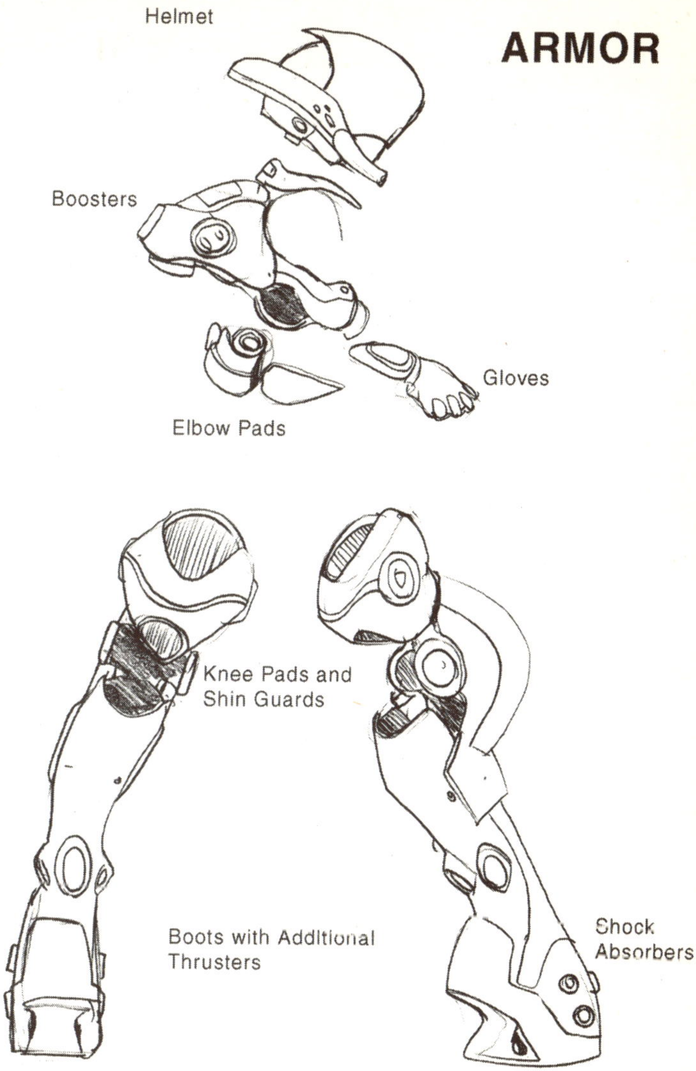
Helmet
Boosters
Gloves
Elbow Pads
Knee Pads and Shin Guards
Boots with Additional Thrusters
Shock Absorbers

These are the basic portions of the armor. It covers up the bare essentials of the user, fitting atop the protective undergarment.

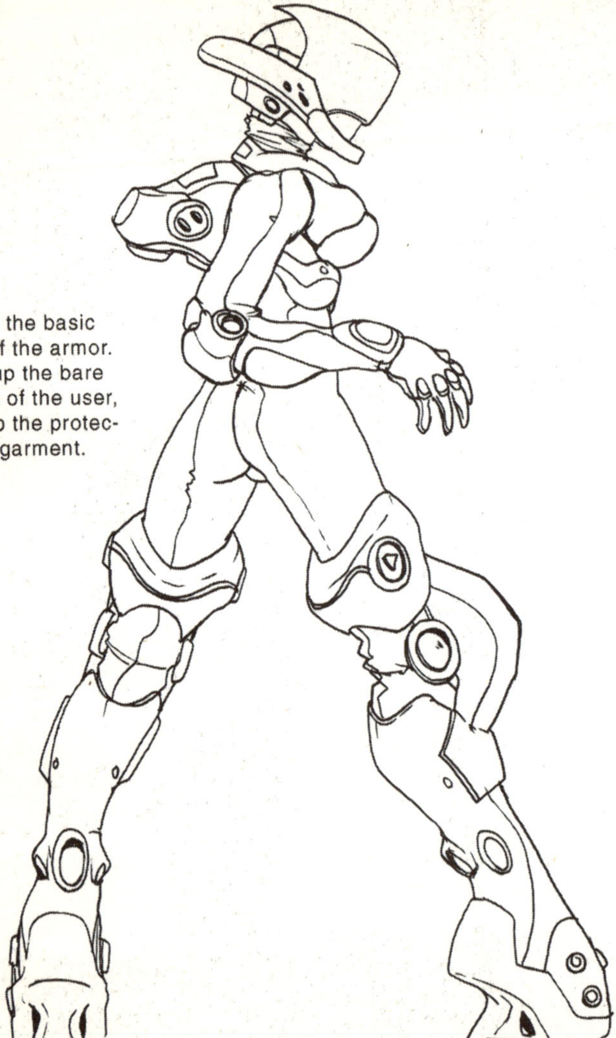

CHAPTER 10

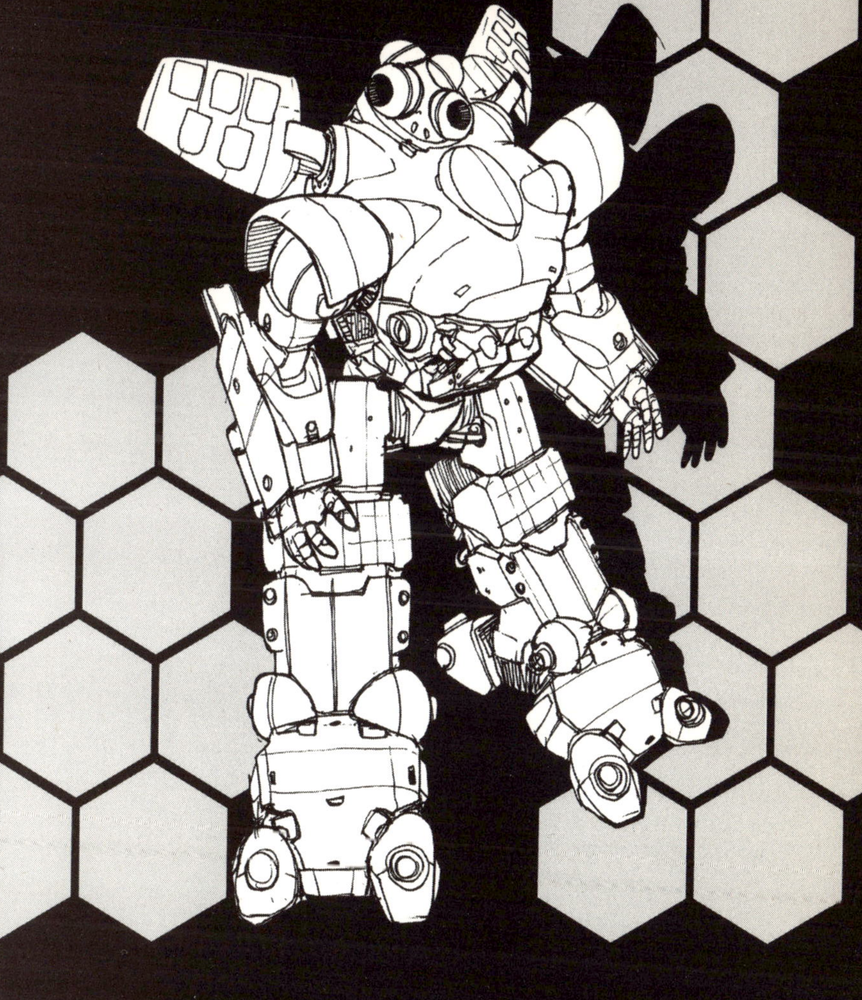

BOOSTERS

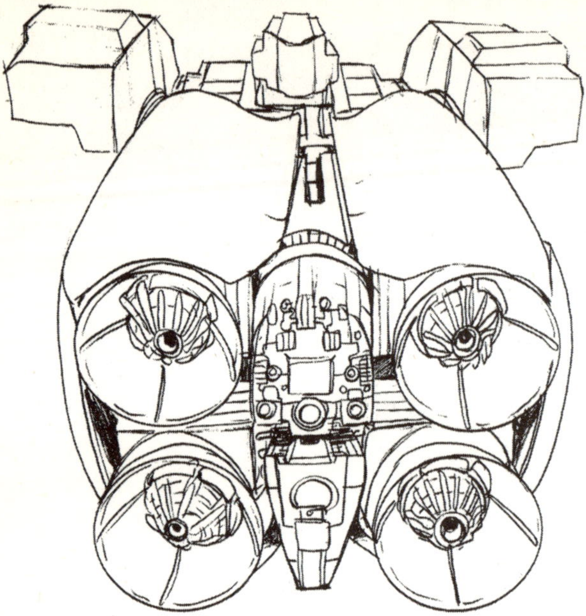

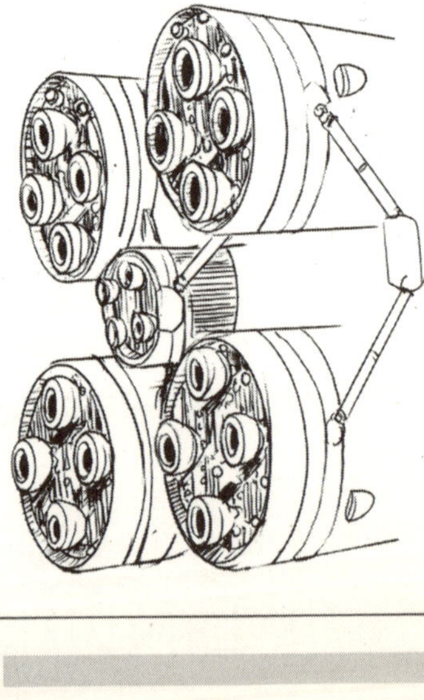

Boosters and rockets can either be incorporated in the design of the mecha or as an external unit. Four rockets are used for this mecha -- or spaceship -- to propel the object at high speeds over great distances. Visually, the larger you make a booster, the more power you're suggesting it has.

BOOSTERS AND RETRO ROCKETS

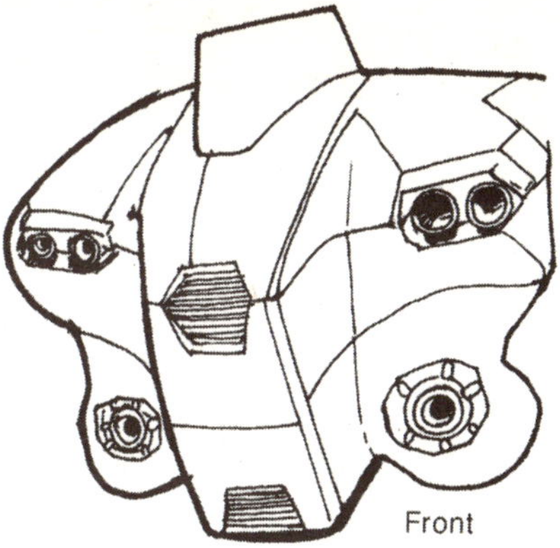

Front

This mecha torso contains forward and rear boosters to control its flight and dampen its inertia. These smaller engines are called retro rockets. The retro rockets are usually incorporated into the mech's design and are not as conspicuous as the larger boosters.

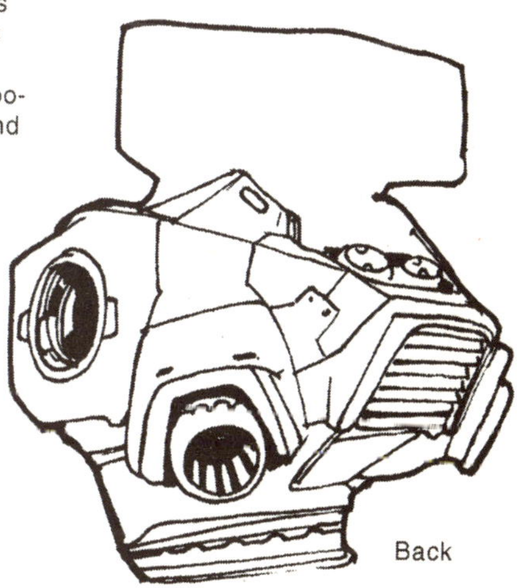

Back

These boosters were made for added speed. When the mech is running, they allow it to cover greater distance in each stride.

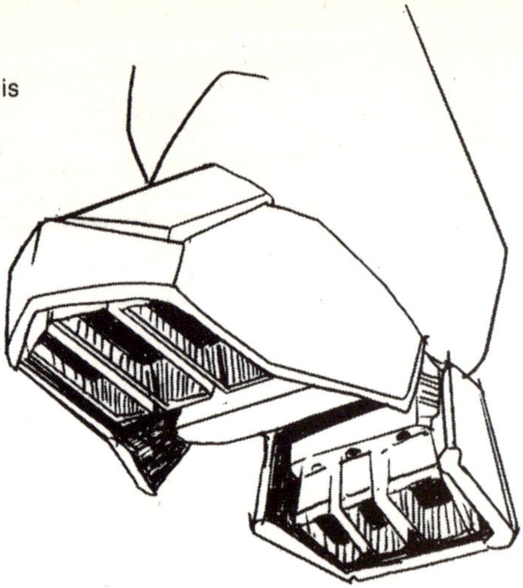

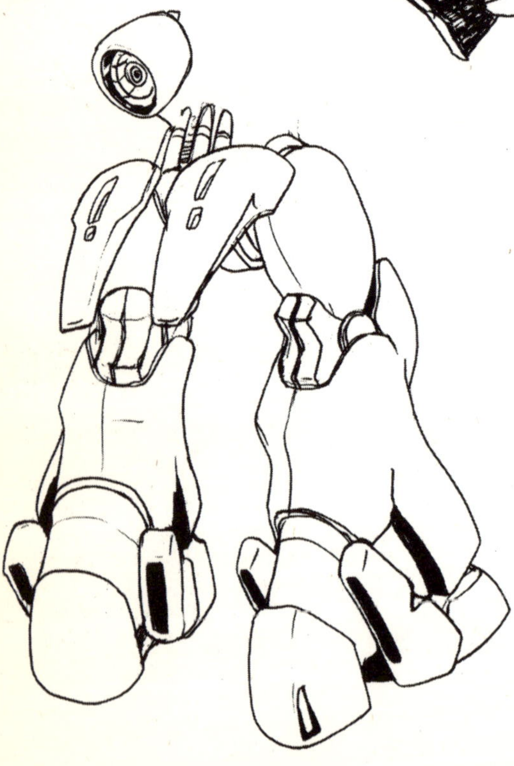

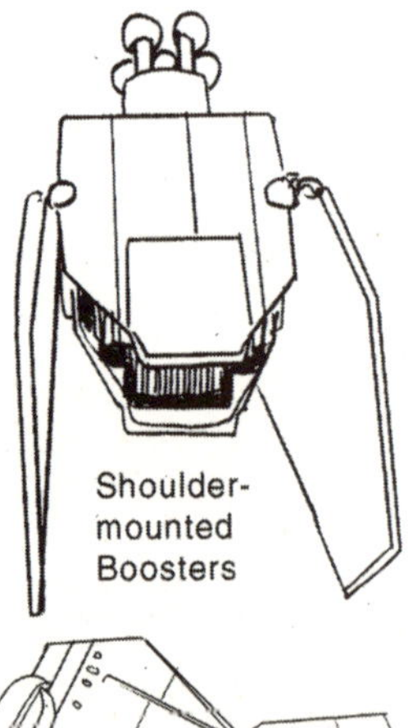

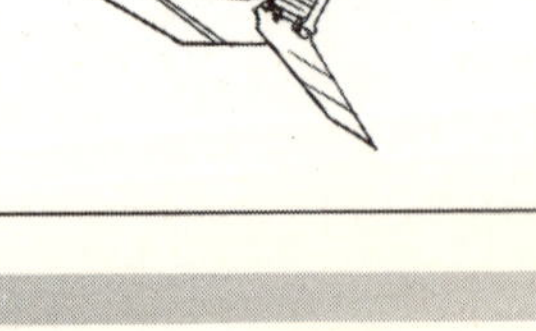

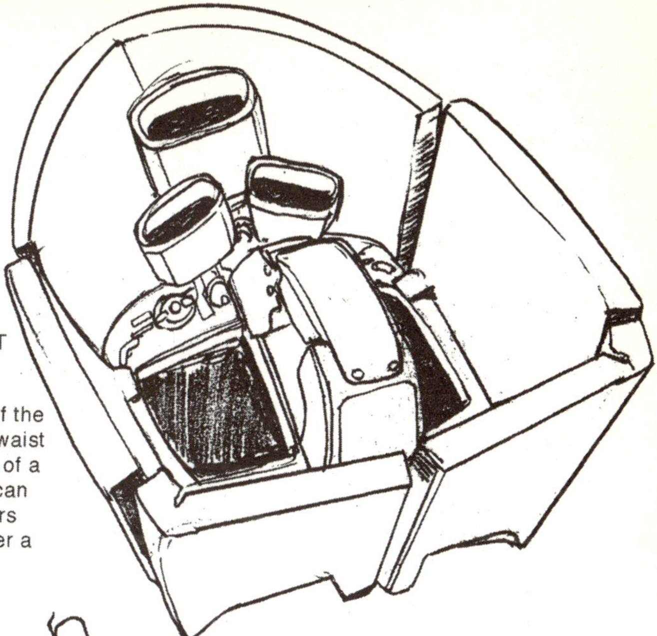

HIP AND WAIST BOOSTERS

This is a view of the underside of a waist and hip section of a mech. As you can see, the boosters are hidden under a housing until needed.

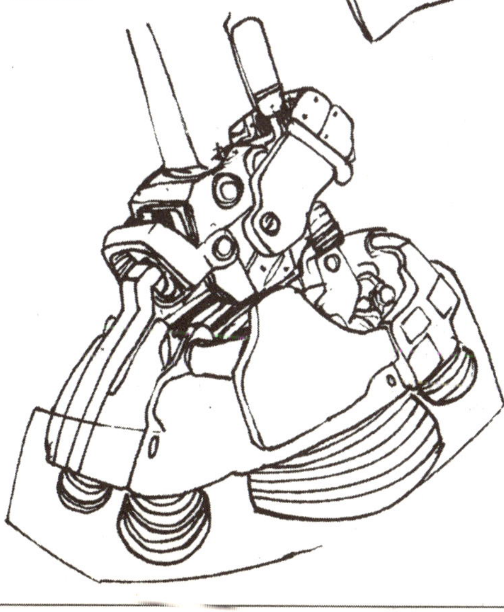

HIDDEN BOOSTERS

FOOT AND LEG BOOSTERS

Here, we have a cut-away of hidden foot boosters. The outside housing of the foot conceals the boosters and protects the engine components, as well as providing the base of the foot for the mech.

MISSILES AND CONTRAILS

Every mech has an arsenal of advanced weapons. A common challenge for artists is to show those weapons doing their thing. Illustrating missiles and the "trails" that they leave is easier than it looks. Just follow this step-by-step!

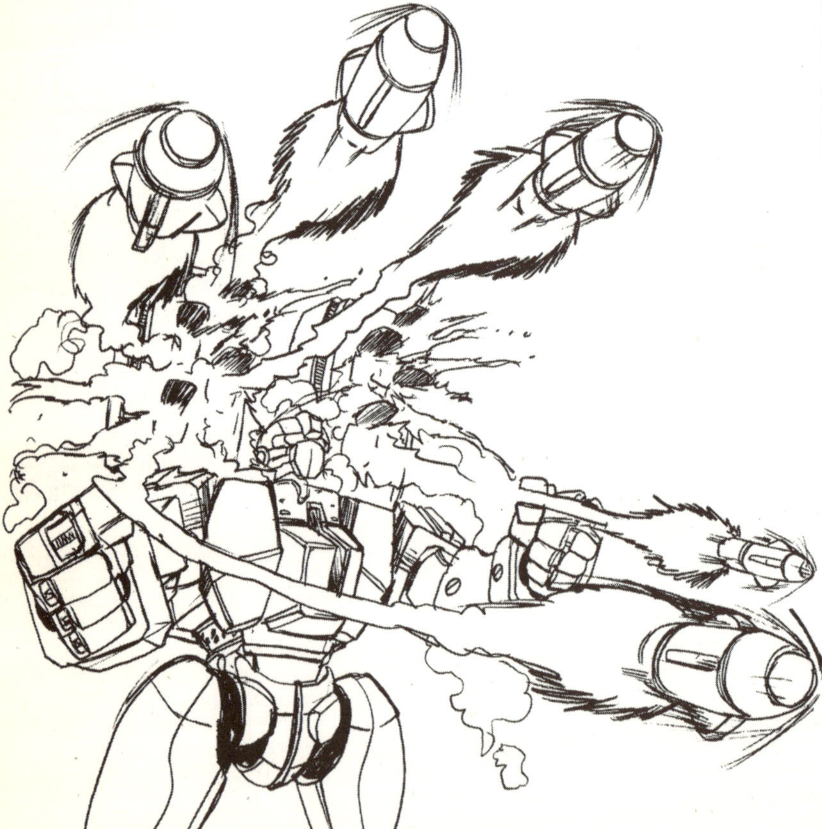

1

Begin by sketching your layout and missile paths using simple lines for the trails and geometric shapes for the form.

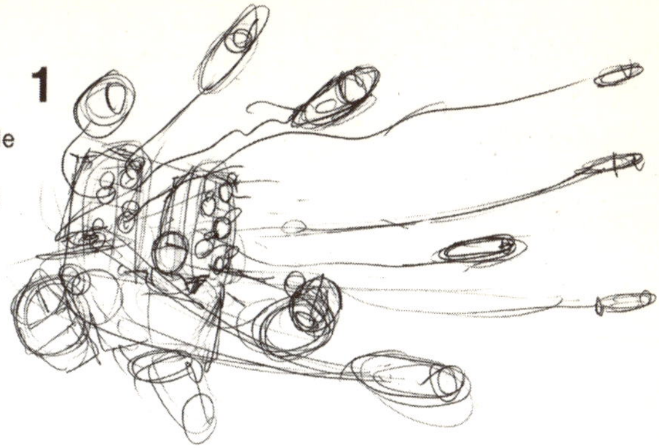

2

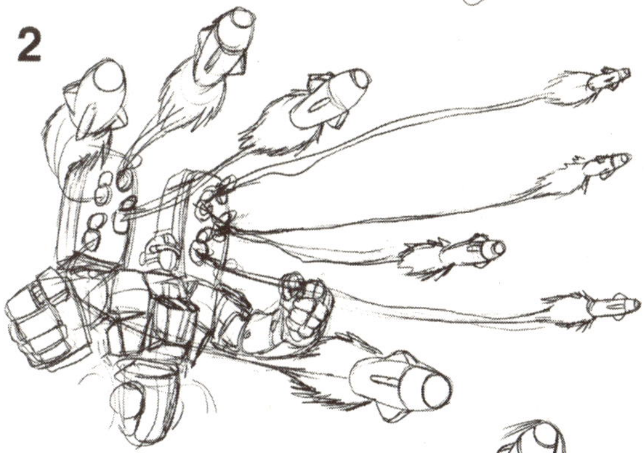

Then, start getting the general details of the form and missiles. Remember that smoke isn't perfect, so you don't have to be either.

3

Finally, fill in all your details. A thing to keep in mind when you're drawing smoke trails is to treat the smoke as one solid shape instead of several wispy shapes. This will make it easier to figure out when it comes time to draw it.

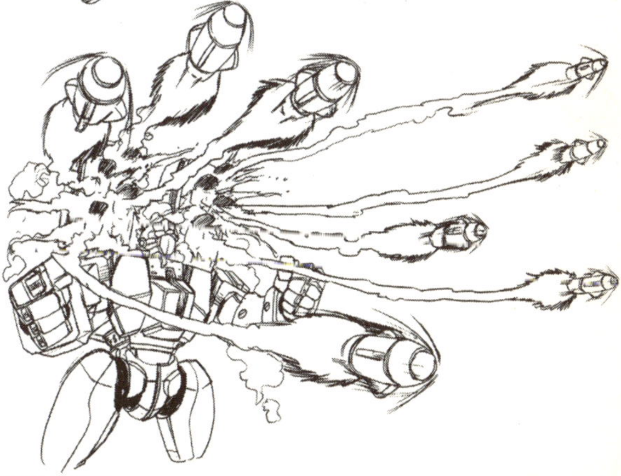

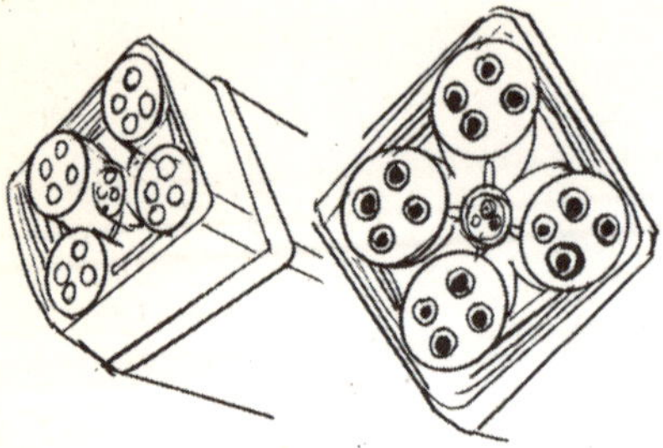

The top boosters are set up in clusters. The below illustration has a set of three fixed boosters.

FIXED BOOSTERS

They're meant to propel the mech forward. With wheel-mounted feet, the mech is propelled horizontally.

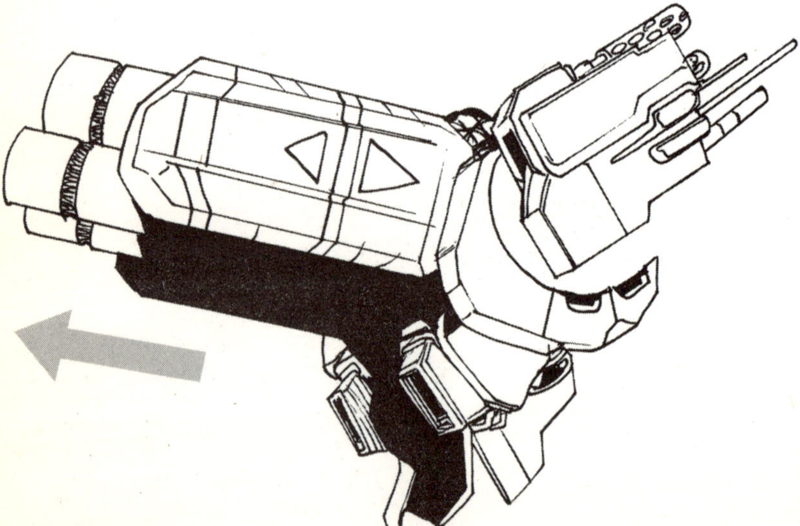

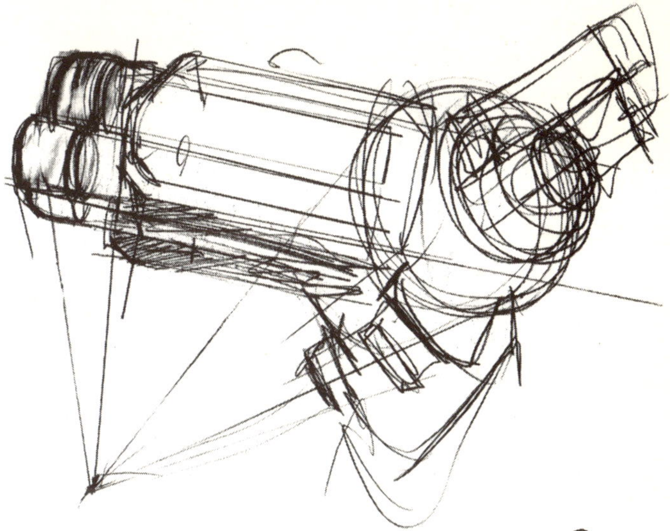

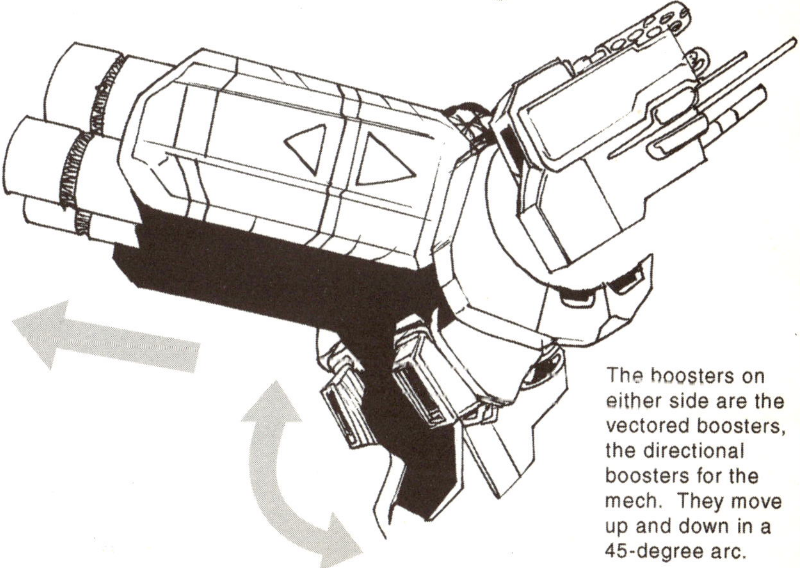

The boosters on either side are the vectored boosters, the directional boosters for the mech. They move up and down in a 45-degree arc.

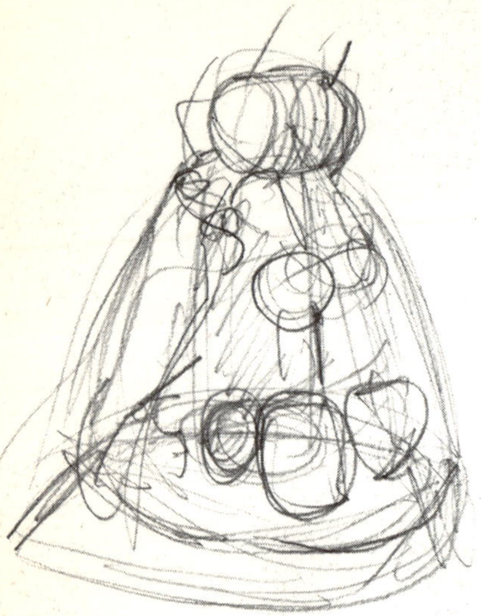

BOOSTER BATTLE DAMAGE

When drawing damage, you want to set up the piece so that the exposed inner workings look realistic and like they would perform a specific function, not like just something that's been thrown together.
It helps to rough-draw the complete interiors so you'll have a better idea of what you need to show.

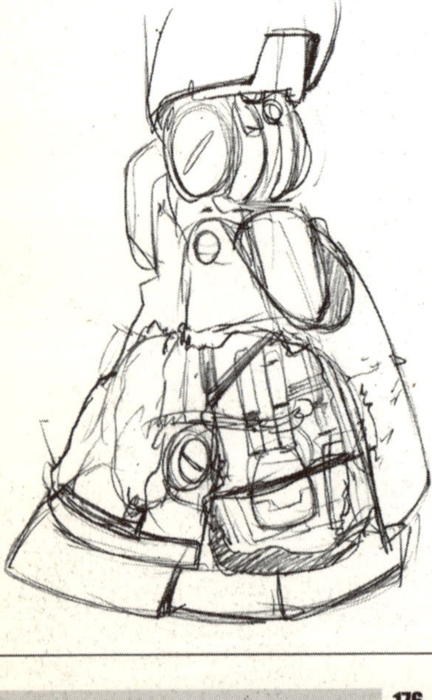

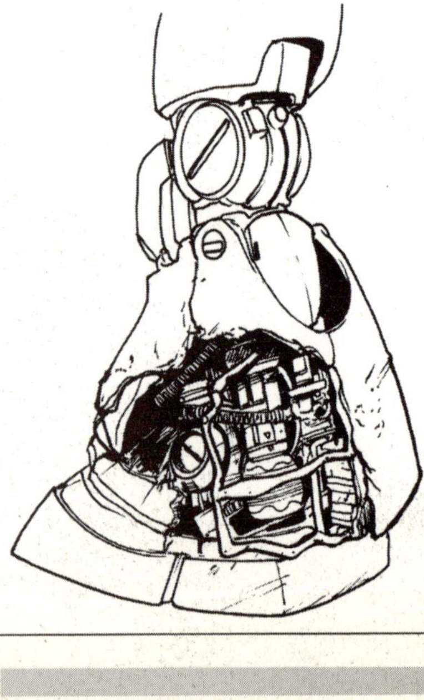

CHAPTER 11

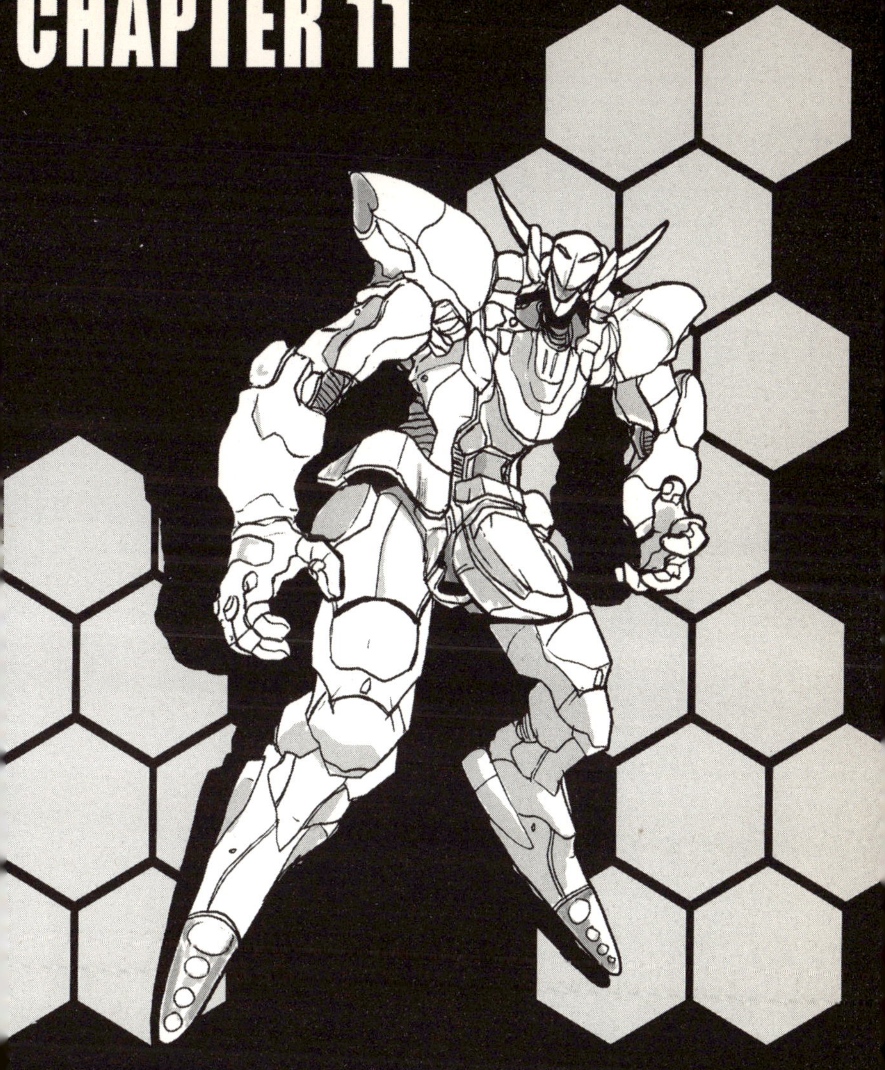

FAKING IT

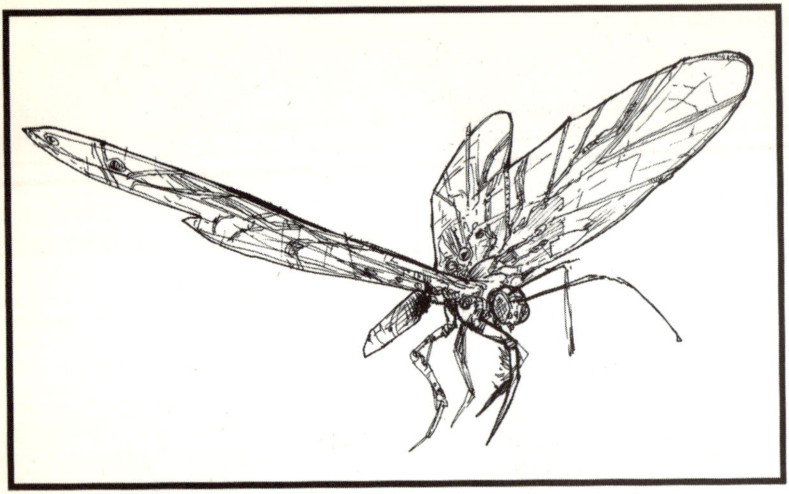

DISCLAIMER

This chapter is meant for entertainment purposes only and is not meant for practical application for anyone studying mech design. However, if you have little or no artistic talent and are looking for a way to get by until you get better, then this chapter's for you!

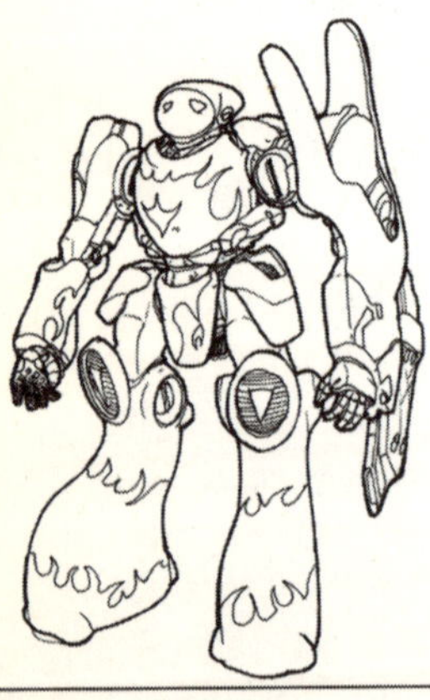

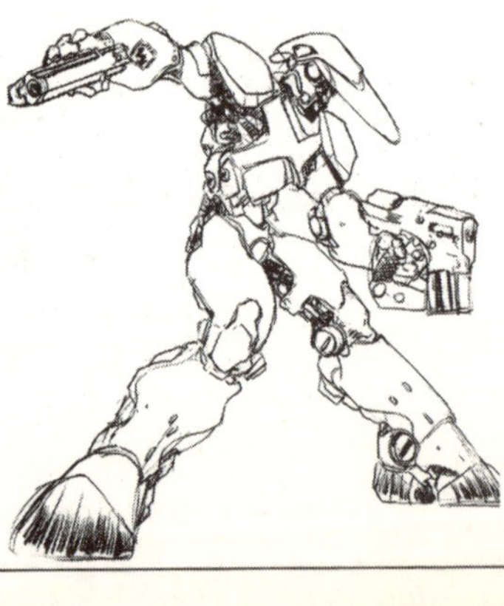

FAKE GUNS

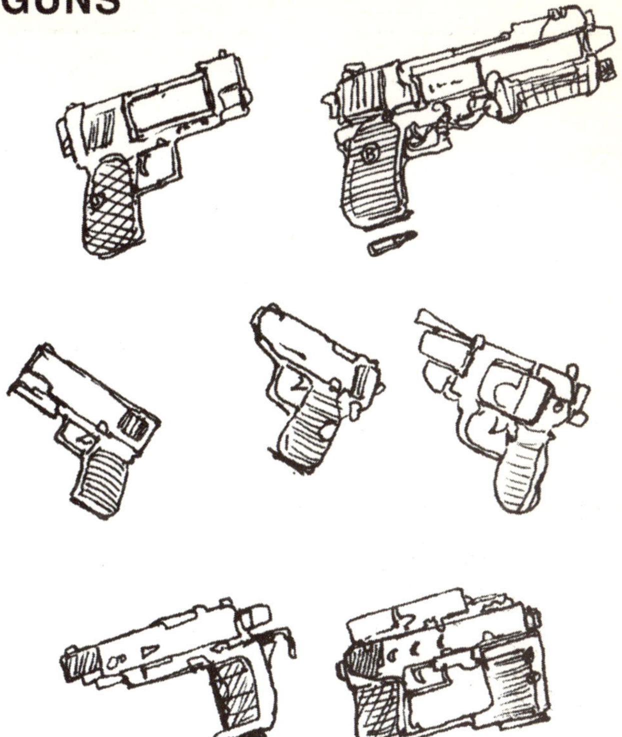

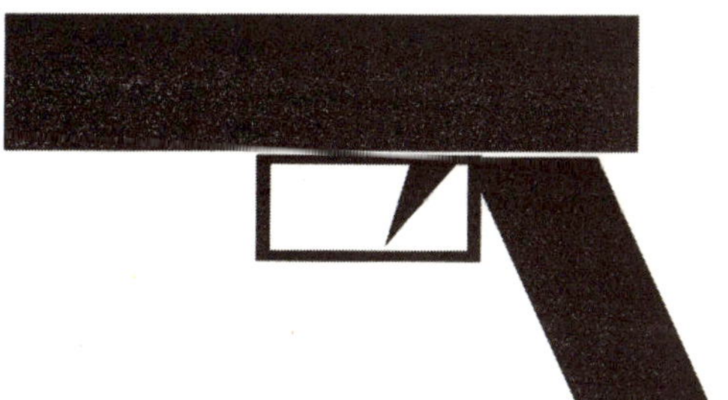

Make up guns, but keep the general structure the same. If you can imply that there's a hand grip, a slide, and a trigger, you can design any type of gun you want.

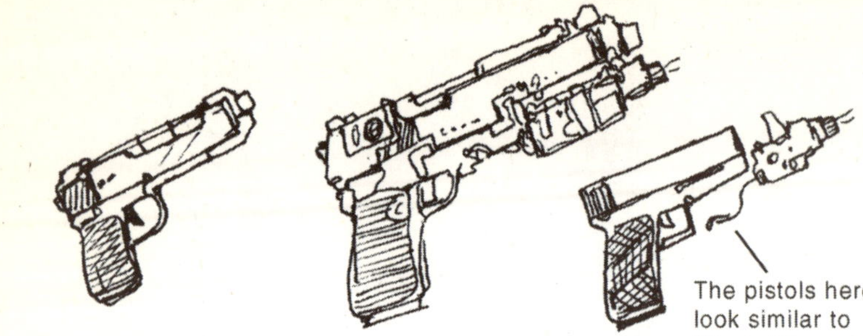

The pistols here look similar to Glock-17s and have pieces that look like add-on laser scopes.

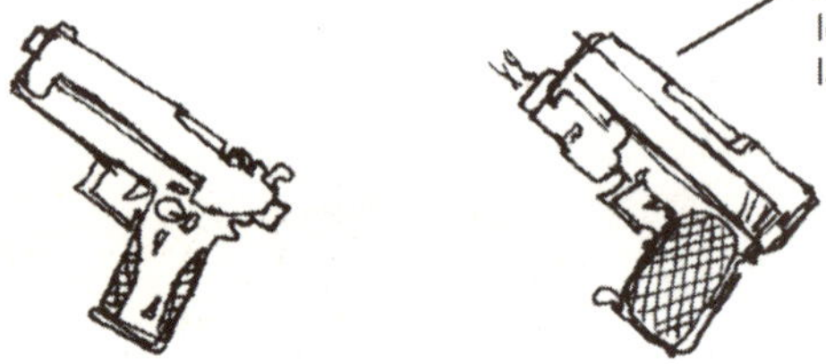

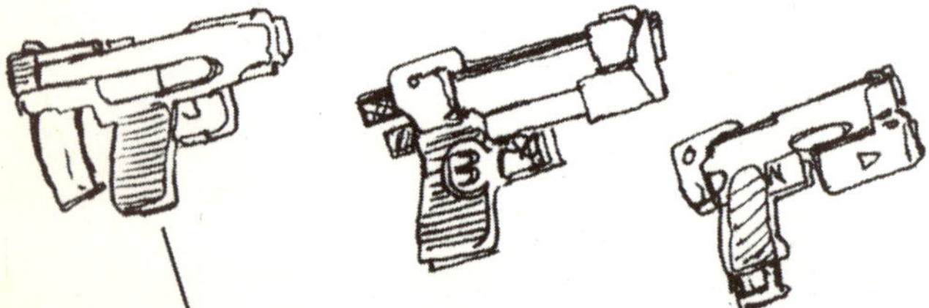

This pistol here has a bullpup design (the magazine is positioned behind the hand grip) that would NOT work in the real world, but it "looks cool."

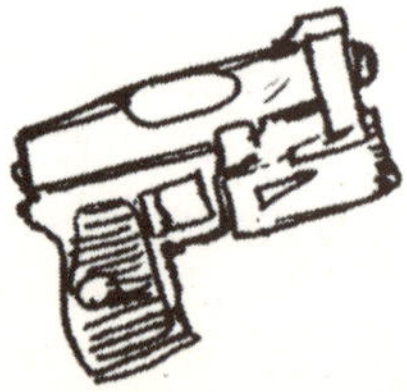

BIKE IN PROFILE

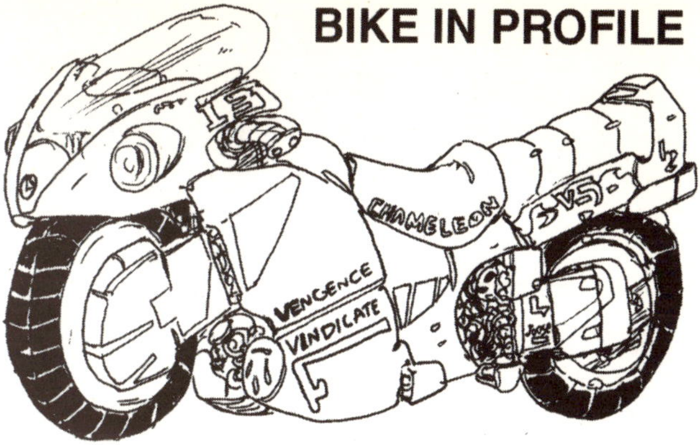

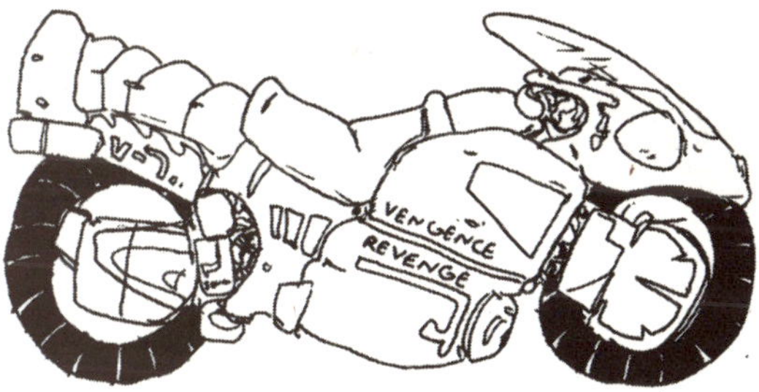

The overly organic design makes it feel like it has personality. The decals on the side give the impression the bike belongs to an over-compensating tough guy.

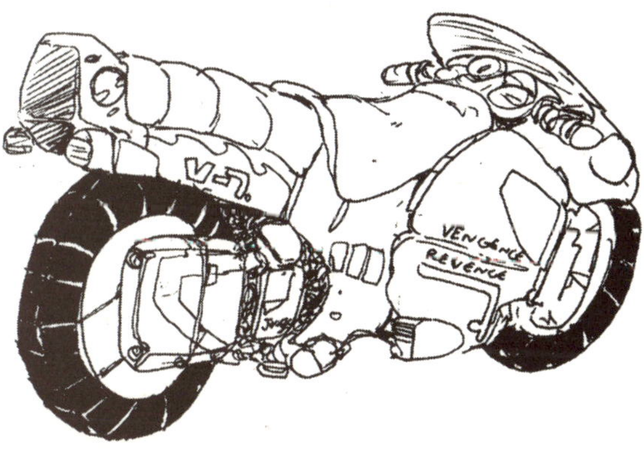

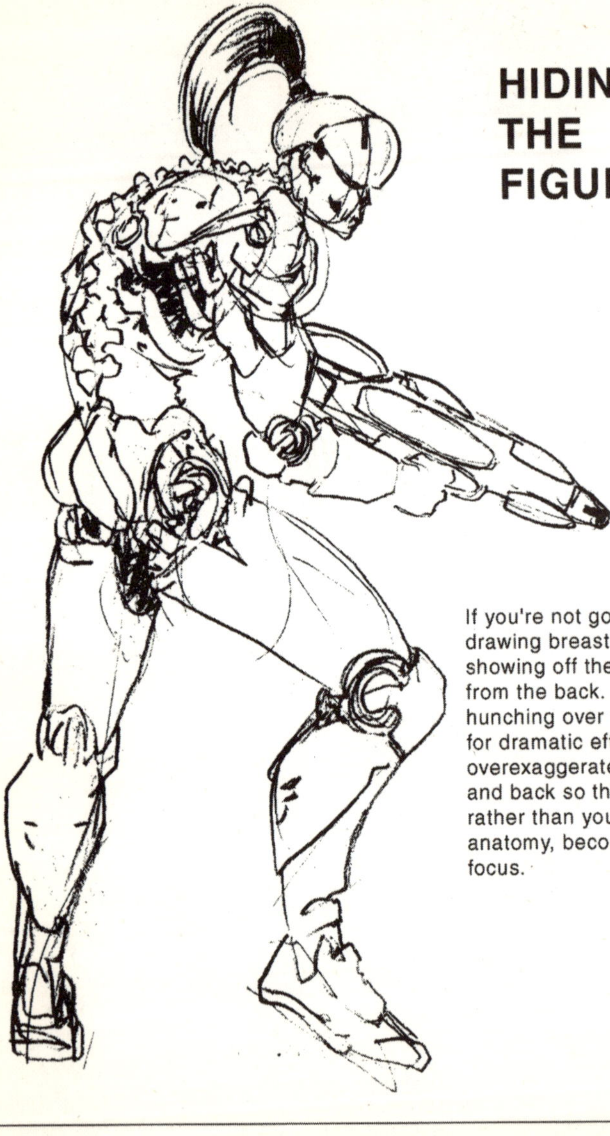

HIDING THE FIGURE

If you're not good at drawing breasts, try showing off the figure from the back. Have it hunching over with a gun for dramatic effect. Also, overexaggerate the spine and back so that area, rather than your lack of anatomy, becomes your focus.

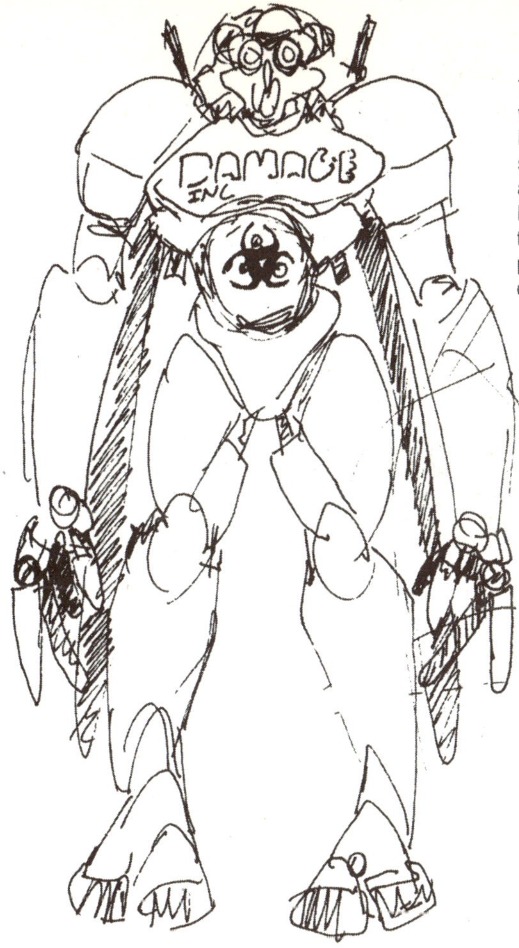

This is supposed to be a worker drone of some kind. Draw it encased in its armor so you don't have to worry about drawing the joints. Add long rods on its back to make the backpack look like a power source or external controlling device.

Add new twists to familiar logos.

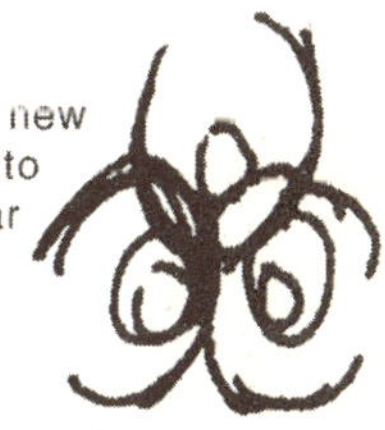

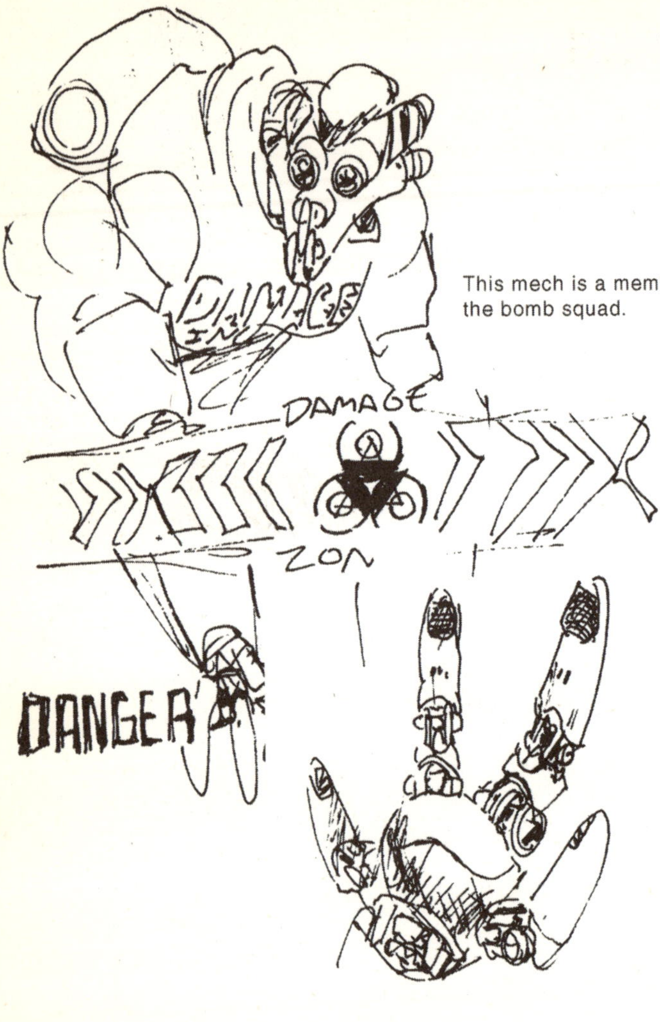

This mech is a member of the bomb squad.

CLUTTER

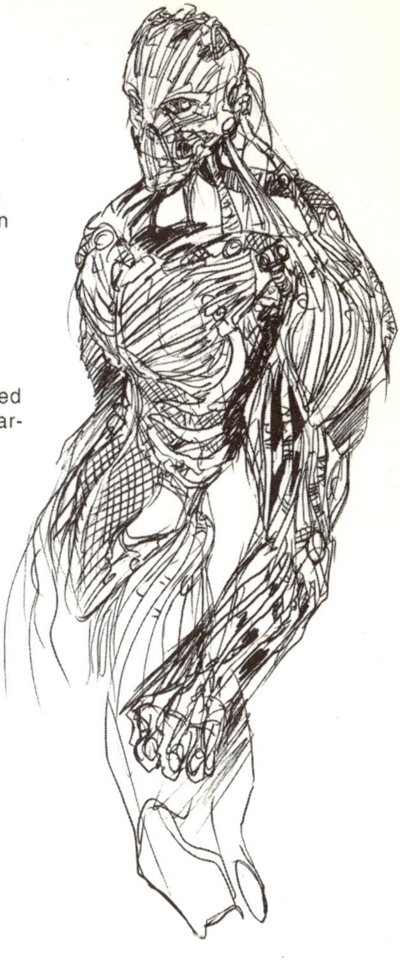

Cluttering up the piece is a good way to make something look more detailed than it actually is. The piece to the right is a series of intertwining lines that lead to nowhere, but they deceptively flow into each other.

The piece below has paneling with no detail sitting in front of an overly rendered interior. This helps you focus on the character's form rather than the inaccurate detail.

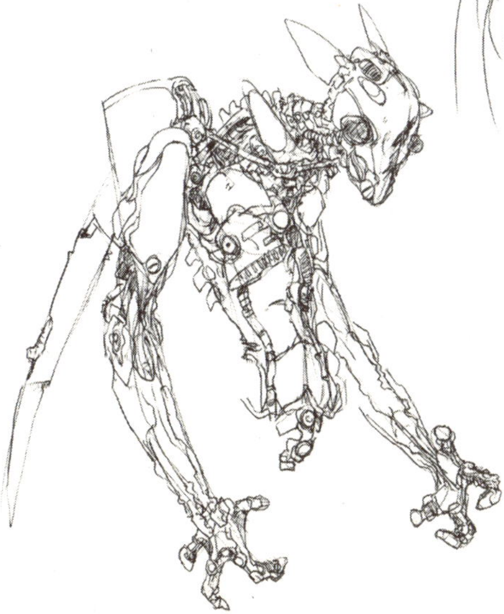

ADVANCED CLUTTER

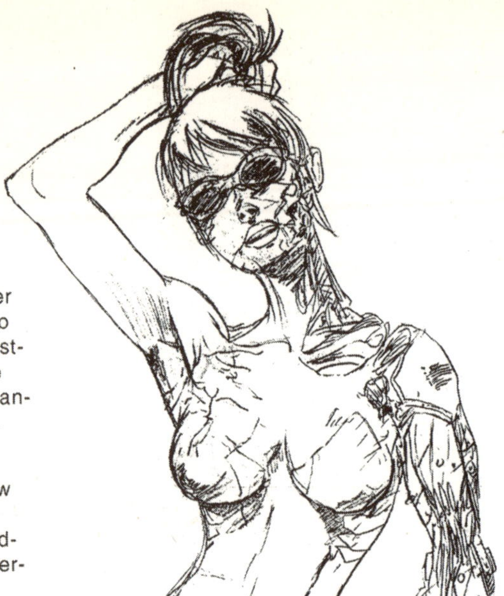

Throwing in spots of clutter on a form is a tricky way to make a bland piece interesting because you focus the viewer's attention on the randomly detailed spots.

The illustrations here show battle-damaged cyborgs. Making the skin look shredded and exposing the cybernetic musculature of the character helps heighten the drama of the piece.

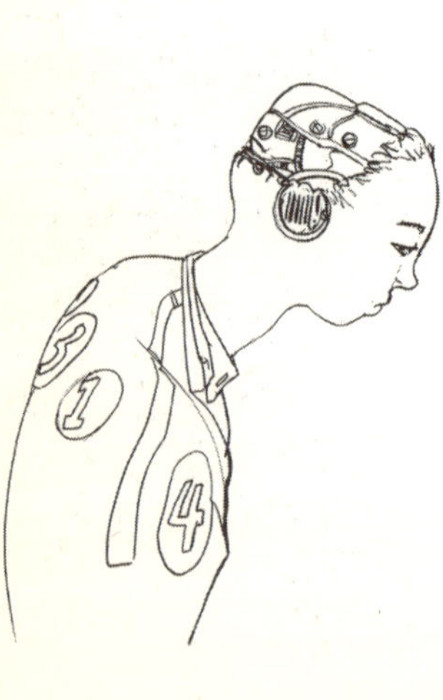

MAKE IT STAND OUT

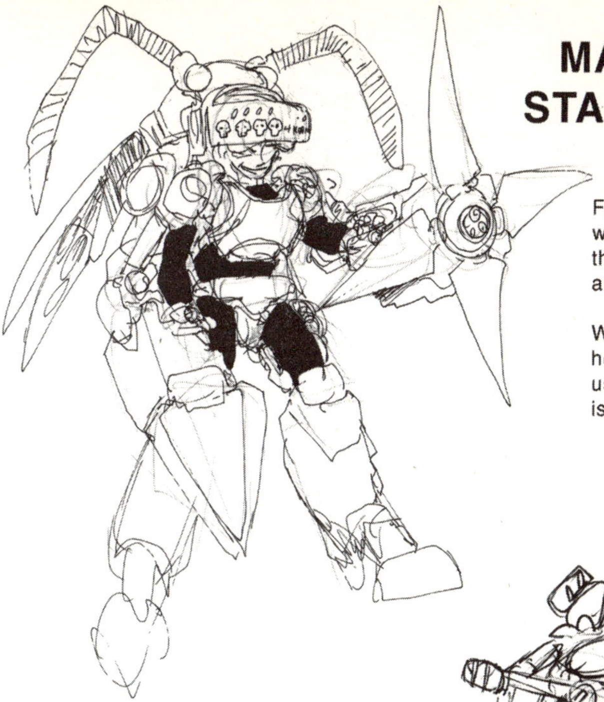

Fill certain areas with black to make the surrounding areas stand out.

When the piece is heavily teched out, using lots of black is a time saver.

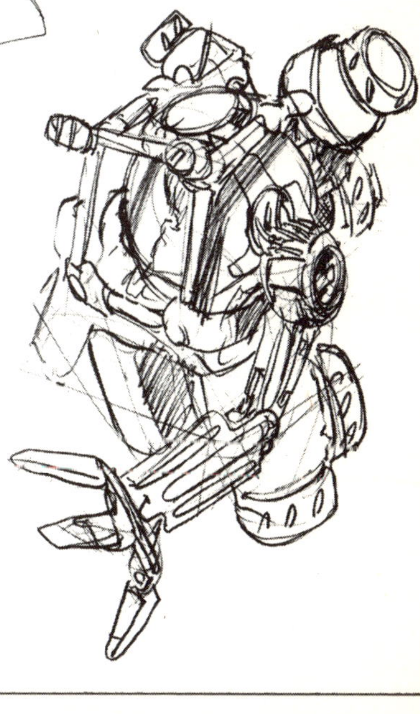

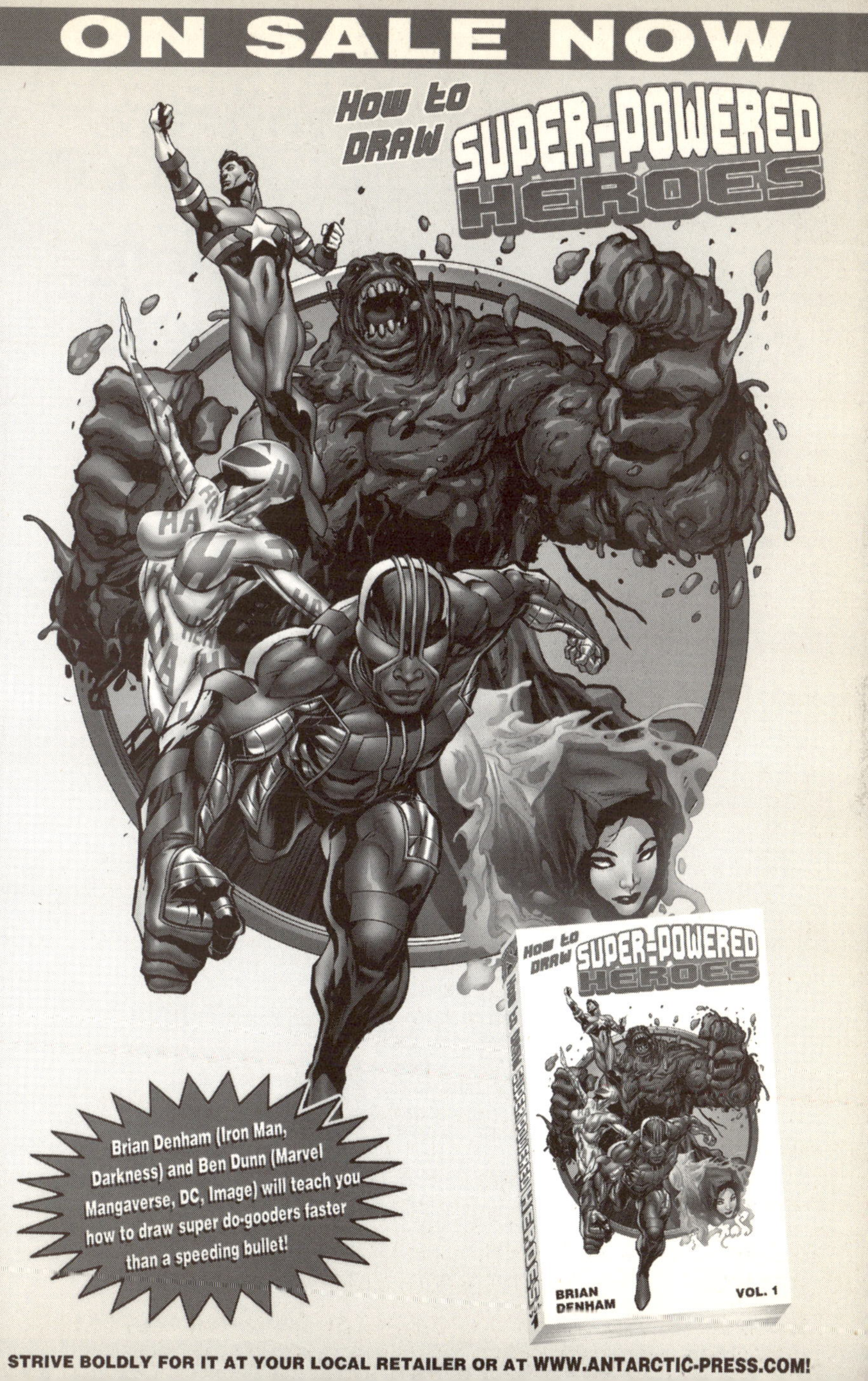
ON SALE NOW
How to DRAW SUPER-POWERED HEROES
Brian Denham (Iron Man, Darkness) and Ben Dunn (Marvel Mangaverse, DC, Image) will teach you how to draw super do-gooders faster than a speeding bullet!
How to DRAW SUPER-POWERED HEROES
BRIAN DENHAM
VOL. 1
STRIVE BOLDLY FOR IT AT YOUR LOCAL RETAILER OR AT WWW.ANTARCTIC-PRESS.COM!

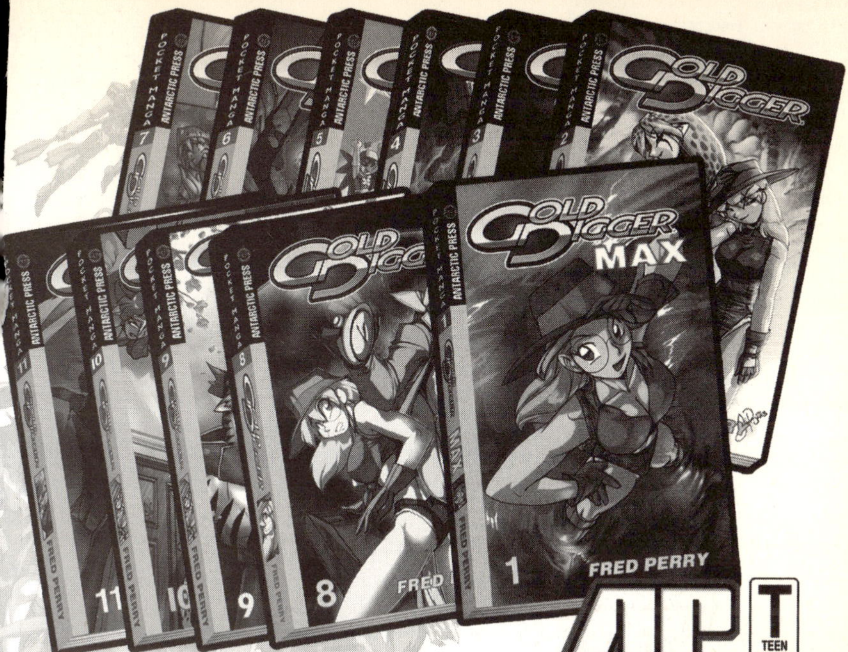

AP T TEEN AGE 13+

POCKET MANGA

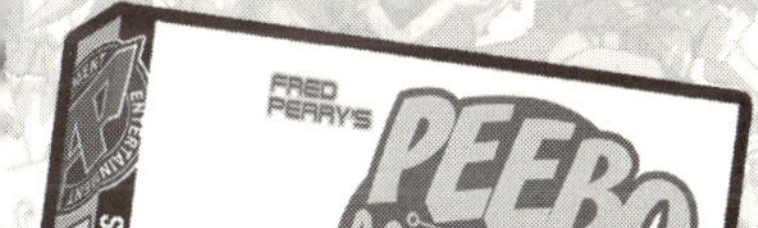

Gold Digger PEEBOmanga 1.0

written and illustrated by
FRED PERRY

Fred Perry drops the bomb on the manga market! Those adorable (super-)smart bombs from Gold Digger are blasting off in their own book! These pint-sized powderkegs score hit after hit with high-explosive humor, rounds of gags sure to pierce the funny bone of any reader! Order your copy of this ALL-NEW, original graphic novel today, before they all go BOOM!

AP T TEEN AGE 13+

POCKET MANGA

POCKETMANGA
Trade Paperback
BLACK AND WHITE
5" x 7"
144 pp
AGE: 13+ teen
Manga/Adventure/Humor
ISBN: 1-932453-91-1
PRICE: $9.99

AND BE SURE TO CHECK OUT
FRED'S GD TANGENT WEB COMICS!
HTTP://GD-TANGENT.TSUNAMI-ART.COM

www.antarctic-press.com